Introduction to Telecommunications Systems

P H Smale

Formerly Principal Lecturer
Coventry Technical College

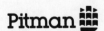

PITMAN PUBLISHING
128 Long Acre, London WC2E 9AN

A Division of Longman Group UK Limited

© P H Smale 1986

First published in Great Britain 1986
Reprinted 1987, 1988, 1989, 1990

ISBN 0 273 02442 6

Printed and bound in Singapore

Contents

1 Introduction to Telecommunications

Introduction

The development of civilization as we know it today is largely due to the human ability to exchange information and ideas by the natural senses of sight and hearing, and by the written word using some form of accepted language or code. From the very beginning, people have constantly searched for means of passing information beyond the normal range of human vision and hearing. Everyone is familiar with such methods as Indian smoke signals, beacon fires, and semaphore flag signalling.

It is worth pointing out here that "tele" is derived from the ancient Greek for "at a distance," "phon" means sound or speech, "graph" means writing or drawing. So the following well-known terms have emerged:

Telecommunication – communicating at a distance.
Telephone – speaking at a distance.
Television – seeing at a distance.
Telegraph – writing at a distance.

Telecommunication is, then, the process of passing **information energy** over long distances by electrical means. The information energy is passed to the destination either over suitable insulated conducting wires called **transmission lines**, or through the atmosphere without the use of wires by a **radio** link. These two methods will be explained later on.

The straightforward use of electrical energy or electricity for everyday tasks is well known; for example electric cookers, electric lights, electric motors, etc. In each of these, electrical energy is converted into another form of energy in order to provide the power to carry out a particular task.

In telecommunication, some form of "information" or "intelligence" energy is changed into electrical energy so that it can be passed to a distant point. At the destination the electrical energy is converted back into its original form. This particular use of electrical energy to convey *information* comes under the general heading of *electronics*. Familiar forms of original information energy are human voice sounds, music, visible moving scenes, still (or non-moving) pictures, and so on.

Basic Requirements of Telecommunications Systems

First of all the original information energy must be converted into electrical form to produce an electronic information **signal**. This is achieved by a suitable **transducer**, which is a general term given to any device that converts energy from one form to another when required.

Suppose the electronic signal is now passed to the destination by a *line link*, with the energy travelling at a speed approaching that of light, and at the destination a second transducer converts the electronic signal back into the original energy form, as shown in Fig. 1.1. In practical systems other items will be required. For example, **amplifiers** may be needed at appropriate points in the system to increase the strength of the electronic signal to acceptable values.

Fig. 1.1 Basic requirements of a one-way line telecommunication channel

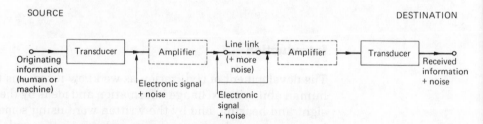

For a *radio* system, a **transmitter** is required at the source to send the signal over the radio link without wires, with the energy travelling at the speed of light, and a **receiver** is required at the destination to recover the signal before applying it to the transducer, as shown in Fig. 1.2.

At this point it is important to realise that, in both these systems, interference will be generated by electronic **noise**, and also that **distortion** of the electronic signal will occur for a number of reasons. These are undesirable effects and must be minimized in the system design.

Fig. 1.2 Basic requirements of a one-way radio telecommunication channel

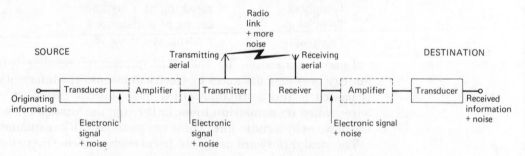

It will be obvious from Fig. 1.1 and Fig. 1.2 that these simple systems are one-way or **unidirectional** only (generally called a **channel**), and domestic radio and television broadcasting are familiar examples of such systems.

Other systems, however, such as the national telephone system, must be capable of conveying information in *both* directions. To do this, the basic requirements shown in Fig. 1.1 and Fig. 1.2 must be duplicated in the opposite direction to provide a two-way or **bothway** system (generally called a **circuit**).

Analog and Coded Signals

Some telecommunication transducers produce an electronic signal that directly follows the instantaneous variations of the original information energy. Such signals are called **analog signals**. For example, a *microphone* produces an electronic signal that follows the variations of sound energy which actuate the microphone. A loudspeaker receives the analog elecronic signal and reproduces the original sound energy variations.

There are other systems in which the transducer produces an electronic signal in the form of a pre-determined **code** of pulses or variations that is understood by humans or machines at both ends of the system. One example of this is the teleprinter, which produces a coded electronic signal dependent on which key is depressed on the sending keyboard. The coded signal is passed to the destination where it is accepted by the receiving teleprinter and the appropriate letter or figure is then printed.

Direct and Alternating Currents

In certain electrical circuits the current flows only in *one* direction when the energy supply is connected, although the amount or strength of the current can be controlled. This type of current is known as **direct current** (d.c.), and is produced by an energy source such as a dry battery, accumulator or rotating generator.

In other electrical circuits, however, the current reverses direction at regular intervals with a particular repeating pattern or **waveform**. This type of current is known as **alternating current** (a.c.), and is generated by an energy source such as a rotating alternator, an electronic oscillator, or certain types of telecommunication transducer.

Use of Direct Current Signals

A steady direct current flowing in a circuit cannot convey information by itself, but the inclusion of a simple on-off switch enables the current to be regulated in a series of pulses. When the switch is opened, the current drops to zero, and when the switch is closed, the current rises to a steady value. If the current pulses are produced in accordance with a pre-arranged code, whereby each letter or number is represented by a particular combination of pulses, then operation of the switch can send any desired message. The current pulses must actuate a device that enables a second person to "see" or "hear" this message.

The Morse Code is one well-known example of this form of d.c. signalling, and a very simple circuit is shown in Fig. 1.3. Each letter has a code

Fig. 1.3 Simple lamp-signalling morse code circuit

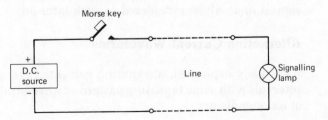

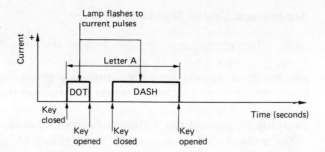

Fig. 1.4 D.C. code representing the letter A in morse code

comprising a fixed number of short and long pulses of current called "dots" and "dashes". The complete alphabet code will not be given here, since it is readily available elsewhere but, for illustration, the combination of current pulses representing the letter A is shown in Fig. 1.4.

If in Fig. 1.3 the connections to the d.c. source are reversed, then the direction of current is also reversed. The current direction can therefore be considered as positive or negative, according to which way it is flowing around the circuit, and this is known as the **polarity** of the current.

With this method of d.c. signalling, the information is carried by the alternate absence and presence of the current. It is also possible to convey information by switching the d.c. between two different values. In either case, it is the variation of *amplitude* of the current that is important.

Other typical uses of d.c. signals are

1 Operating automatic exchange switches from telephone dials.
2 Control and metering of telephone calls between exchanges.
3 Operating simple indicating instruments, such as car fuel gauges, etc.

It should be added here that other systems such as p.c.m., data, and ceefax use d.c. signals in various forms, either in the on/off condition, or using current reversals, but in general these will be considered later as a.c. signals.

The main disadvantages of d.c. signals are

1 Difficulty in transmission over long line circuits due to attenuation (weakening) and distortion, although regeneration (boosting) and amplification are possible.
2 Connecting wires are always needed for the whole of a telecommunication circuit.

It is important to realize, however, that sources providing direct current are widely used to supply power or energy to electronic circuits.

Varying or fluctuating d.c. signals have similar characteristics to a.c. signals and will be considered as such later on.

Alternating Current Waveforms

As already explained, alternating currents reverse direction at regular intervals with some repeating pattern or waveform. The main advantages of a.c. signals are

1 The strength or amplitude can easily be altered (e.g. by transformer, amplifier, etc.), allowing transmission over long lines.
2 Connecting wires are *not* necessarily required for the *whole* of a tele-communication circuit.

Many waveforms are possible with alternating currents. One of the simplest to produce comes by rotation of a loop of wire in a uniform magnetic field. This is called a **sinusoidal waveform** and is shown in Fig. 1.5.

Fig. 1.5 Sinusoidal a.c. waveform

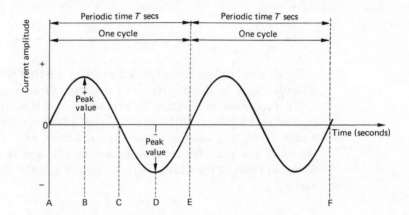

Between points A and B the current increases from zero to a peak value in the positive direction. Between points B and C the current gradually reduces to zero. Then, between points C and D, the current "increases" to a peak value in the opposite or negative direction, and between points D and E it gradually "reduces" to zero again. This whole sequence from point A to point E represents one complete rotation of the wire loop in the magnetic field, and is called one **cycle** of a.c. waveform. Clearly the cycle is repeated between points E and F, representing another rotation of the wire loop, and this waveform is repeated for each subsequent rotation.

The time needed, in seconds, for one cycle of waveform to be produced is called the **periodic time** (*T*) of the a.c. waveform.

The number of complete cycles occuring in one second is called the **frequency** (*f*) of the a.c. waveform in **hertz** (Hz). One hertz is one cycle per second. From Fig. 1.6 it should be clear that frequency and periodic time are reciprocals of each other. That is

$$\text{Frequency} = \frac{1}{\text{Periodic time}} \quad \text{and}$$

$$\text{Periodic time} = \frac{1}{\text{Frequency}}$$

with frequency in hertz (Hz) and time in seconds. From Fig. 1.6, the frequency is 4 Hz and the periodic time is ¼ second.

The strength of the current at any instant in time is called the **amplitude** of the waveform, and the direction of the current (positive or negative) is called the **polarity** of the current, as in d.c.

Fig. 1.6 Sinusoidal a.c. waveform with a frequency of 4 Hz

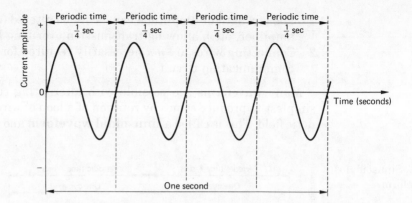

It will also be clear from Fig. 1.5 and Fig. 1.6 that the amplitude reaches a **peak value** in the positive *and* negative directions once every cycle.

We have, as one example, already associated the production of a sinusoidal waveform with the rotation of a loop of wire in a magnetic field, and the resulting current plotted against time, as in Fig. 1.5 and Fig. 1.6. We could also consider the loop as moving through 360° in one rotation, so we could plot the resultant current against angular rotation, as shown in Fig. 1.7.

Fig. 1.7 Sinusoidal a.c. waveform plotted against angular rotation

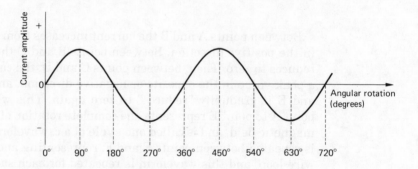

Also, if we consider the energy of an a.c. waveform travelling through space or along a transmission line at a particular velocity, then a certain distance will be travelled in the periodic time for one cycle, as shown in Fig. 1.8. We can see now that a.c. waveform repeats complete cycles over equal distances. The distance representing each cycle is called the **wavelength** of the a.c. waveform in metres. The Greek letter lambda (λ) is used as the symbol for wavelength.

Fig. 1.8 Sinusoidal a.c. waveform plotted against distance

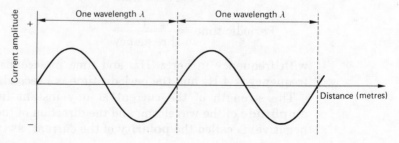

In Fig. 1.7 the rotation of the loop is shown as a continuously increasing number of degrees. Alternatively, we could consider the start of each rotation as beginning from 0°, so each successive cycle in fact occurs from 0 to 360°, as shown in Fig. 1.9. Clearly at the *same point* in each cycle the amplitude of the waveform has the *same value*. This way of identifying a particular point in any cycle as a degree of rotation is called the **phase** of the a.c. waveform.

Fig. 1.9 Sinusoidal a.c. waveform related to phase

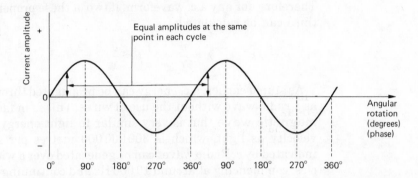

In the same way, if two waveforms are identical *except* for their phase, then the difference between the two can be expressed as a **phase difference**, as shown in Fig. 1.10. In Fig. 1.10, waveform A is seen to be *leading* waveform B by 90°. Put another way, waveform B is *lagging* waveform A by 90°.

Fig. 1.10 Illustration of phase difference ϕ between two waveforms

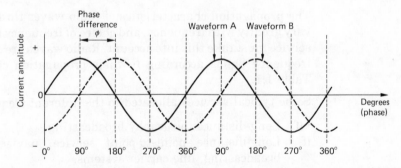

Relationship between Frequency, Wavelength and Velocity

We have seen that an a.c. waveform has a certain energy velocity (metres per second), with a periodic time (T seconds) for the duration of each cycle, and with a certain wavelength distance (λ metres) for each cycle. Now, in general, velocity, distance and time are related by

$$\text{Velocity} = \frac{\text{Distance}}{\text{Time}} \qquad \text{(e.g. metres per second, km/h)}$$

So, for any a.c. waveform of wavelength λ and periodic time T,

$$\text{Velocity } v = \frac{\text{Wavelength } \lambda}{\text{Time } T}$$

Introduction to Telecommunications **7**

But it has already been established that

$$\text{Frequency } f(\text{Hz}) = \frac{1}{\text{Periodic time } T \text{ (secs)}}$$

so that

$$\text{Velocity } v = \text{Wavelength } \lambda \times \text{Frequency } f$$

Therefore, for any a.c. waveform, if two of these properties are known, the third can be calculated:

$$v = \lambda f \qquad \lambda = \frac{v}{f} \qquad f = \frac{v}{\lambda}$$

As stated earlier, a.c. energy can be propagated through the atmosphere as a radio wave without the use of wires. This is in fact a type of **electromagnetic wave** that is very similar to light energy, and has the same velocity as light, which is 300 000 000 metres per second and usually indicated by c. Radio waves can be generated over a wide range of frequencies, commencing at around 10 000 Hz and continuing through millions of hertz to thousands of millions of hertz, as indicated in Fig. 1.11. Included also are electromagnetic waves such as visible light, X-rays, etc.

The following abbreviations should be noted:

thousands of hertz, or kilohertz – kHz
millions of hertz, or megahertz – MHz
thousands of millions of hertz, or gigahertz – GHz

The propagation characteristics of radio waves through the atmosphere vary greatly with frequency, and choice of frequency for a particular radio service must take this into account. Radio waves are divided into different frequency bands according to their propagation characteristics, as in Table 1.1.

Some typical services allocated to the different frequency bands are

vlf Long-distance telegraphy broadcasting.
lf   Long-distance point-to-point service, navigational aids, sound broadcasting, line carrier systems.
mf   Sound broadcasting, ship-shore services, line carrier systems.
hf   Medium and long-distance point-to-point services, sound broadcasting, line carrier systems.
vhf⎫ Short-distance communication, tv and sound broadcasting, radar.
uhf⎭ Air-air and air-ground services.
shf Point-to-point microwave communication systems, radar.

Knowing the *velocity* of radio waves, then when a certain radio transmission is operating at an allocated *frequency*, the corresponding *wavelength* can be calculated.

EXAMPLE 1.1
The BBC Radio 2 sound programme is received by domestic radio receivers on 1500 metres in the Long Waveband. What is the allocated frequency of the programme?

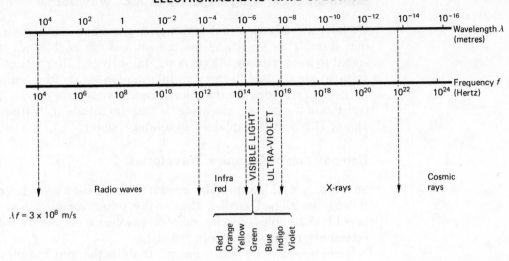

Fig. 1.11 ELECTROMAGNETIC WAVE SPECTRUM

$\lambda f = 3 \times 10^8$ m/s

Table 1.1

Frequency band		Corresponding wavelength
Very low frequency (vlf)	below 30 kHz	above 10 000 metres
Low frequency (lf)	30 kHz to 300 kHz	10 000 to 1000 m
Medium frequency (mf)	300 kHz to 3 MHz	1000 to 100 m
High frequency (hf)	3 MHz to 30 MHz	100 to 10 m
Very high frequency (vhf)	30 MHz to 300 MHz	10 to 1 m
Ultra high frequency (uhf)	300 MHz to 3 GHz	1 m to 10 cm
Super high frequency (shf)	3 GHz to 30 GHz	10 cm to 1 cm
Extremely high frequency (ehf)	above 30 GHz	below 1 cm

Now, Velocity c = 300 million metres per second, and Wavelength λ = 1500 metres.
Therefore

$$\text{Frequency } f = \frac{\text{Velocity } c}{\text{Wavelength } \lambda} = \frac{300\,000\,000}{1500} \text{ Hz}$$

$$= 200\,000 \text{ Hz} \quad \text{or} \quad 200 \text{ kHz} \quad (Ans.)$$

EXAMPLE 1.2
If the frequency allocated to Radio Luxembourg is 1.442307 MHz, calculate the corresponding wavelength.

$$\text{Wavelength } \lambda = \frac{\text{Velocity } c}{\text{Frequency (Hz)}}$$

$$= \frac{300\,000\,000}{1\,442\,307} \text{ metres}$$

$$= 208 \text{ metres} \quad (Ans.)$$

Information-carrying Capacity of A.C. Waveforms

We have already seen that variation in the amplitude of a direct current (for example by switching the d.c. on and off) enables an information signal to be carried by the current. Similarly, an alternating current (or voltage) can be used to carry an information signal from source to destination by arranging for the information signal to vary one of the characteristics of the a.c. waveform – either its amplitude or its frequency or its phase. This will be explained in a later chapter.

Composition of Complex Waveforms

So far, only simple sinusoidal waveforms have been considered, although it was mentioned earlier that many other complex waveforms are possible. Examples of some complex waveforms commonly found in telecommunication are shown in Fig. 1.12.

It can be shown by mathematical analysis that any complex waveform is made up of a sinusoidal waveform having a certain frequency called the *fundamental* frequency and a number of other sinusoidal waveforms having frequencies that are direct multiples of the fundamental frequency with decreasing peak values. These direct multiples are called **harmonics** of the fundamental frequency. Some other complex waveforms will contain a d.c. component also, although this is zero for the three waveforms shown in Fig. 1.12.

For example, for a complex waveform having a fundamental frequency f Hz the following harmonics may also be present:

$2f, 3f, 4f, 5f, 6f, \ldots$ etc.

The square wave shown in Fig. 1.12a is made up theoretically of a fundamental frequency f and all the **odd** harmonics rising to infinity, i.e.

$3f, 5f, 7f, \ldots$ etc.

Fig. 1.13 shows how a sinuisoidal fundamental waveform and its 3rd and 5th harmonics combine to produce a complex wave, which suggests that a square wave would be produced if further odd harmonics were added.

The saw-tooth waveform shown in Fig. 1.12b contains a fundamental sinusoidal waveform and, theoretically, all odd and even harmonics to infinity.

Information Signal Bandwidth

The sound waves produced by a human voice are variations of air pressure above and below normal pressure, so they can be considered as alternating in nature, with a complex waveform that is different for each individual voice. It is this unique nature of the complex waveform of sound waves that enables us to recognize individual voices. Since each voice has a different complex waveform, it must contain certain fundamental frequencies and harmonics. Generally, the range of fundamental frequencies represents the information or intelligence, whilst the harmonic content

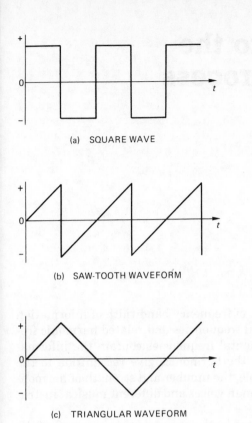

(a)   SQUARE WAVE

(b)   SAW-TOOTH WAVEFORM

(c)   TRIANGULAR WAVEFORM

Fig. 1.12   Typical
complex a.c.
waveforms

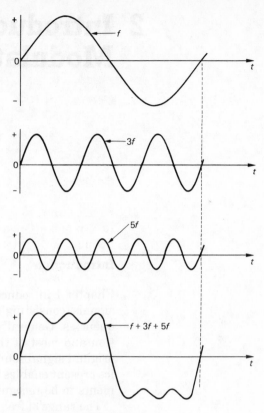

Fig. 1.13   Amplitude
addition of fundamental a.c.
and its 3rd and 5th harmonics

gives individual recognition. The sound waves produced by human voices
therefore must contain a **range** of frequencies, and the range is known as
the **bandwidth**. Since the microphone produces an electronic signal that
is virtually the direct analog or copy of the sound waves, the speech information electronic signal must also have a minimum frequency
bandwidth that must be maintained throughout the system carrying the
information.

Other different types of information signal in telecommunication (telegraphy, television, music, data, etc.) have different minimum
bandwidths, which will be considered later.

It should be realized that this idea of information signal bandwidth has
an important bearing on the design of a system, in terms of circuitry and
transmission media.

2 Introduction to the Modulation Process

Introduction

Chapter 1 introduced the concept of frequency bandwidth of information signals, consisting of fundamental frequencies and related harmonic frequencies. Generally, the fundamental frequencies contain the information and most of the power, and the harmonics give recognition to the original signal source. For example, the number and strength of harmonics present enables individual human voices and different musical instruments to be recognized.

The range of frequencies produced by the average human voice is of the order of 100–7500 Hz. As a simple rule, it can be said that frequency bandwidth transmission costs money, so at a very early stage in the development of national and international telephone networks it was agreed that, in the interests of economy, the natural frequency bandwidth of speech signals produced by talking into a telephone should be restricted to the range 300–3400 Hz, called the **commercial speech bandwidth**. This enables the information to be readily understood by the listener at the distant end without necessarily giving recognition of the talker's voice. In fact, however, it is generally found that the "telephone" voice of any individual becomes recognizable after relatively few conversations.

The range of frequencies produced by a full musical orchestra can be of the order of 30–20 000 Hz, which also is the approximate hearing range of most people with good hearing facilities. For aged people and those with certain hearing defects, the range of frequencies that can be detected will probably be less than this.

In order to enjoy music fully it is essential that the individuality of instruments is not destroyed, so it is not advisable to restrict the natural frequency bandwidth produced. The frequency bandwidth used for music transmission depends on the particular situation, but some common examples are

a   For medium wave radio broadcasting and associated land-links 50–8000 Hz.
b   For BBC vhf/fm sound broadcasting, 50–15 000 Hz.
c   For high-fidelity sound reproduction (hi-fi), 30–20 000 Hz.

There are two different situations which need the introduction of the modulation process:

1 At certain points in the national telephone network, the number of telephone calls being handled is very large, and if a separate line or radio bearer were used for every call having a frequency bandwidth of 300–3400 Hz, then very many such bearers would be necessary. It is possible to manufacture cables or install radio links which can handle signals over very large frequency bandwidths, so it was a natural development to look for ways of allowing many telephone calls to *share* the large bandwidth capacity of these bearers.

This technique is called **multiplexing**, and the sharing of a line by many telephone channels can be done either on a frequency basis called *frequency-division multiplex*, or on a time basis called *time-division multiplex*. The principles involved are considered in Chapter 10.

2 It was seen in Chapter 1 that it is possible to radiate energy as an electromagnetic wave without the use of wires at frequencies from about 10 kHz upwards. In fact, at very low frequencies (under 30 kHz) it is very expensive to transmit radio waves because of the high power needed from the transmitter, and because the transmitter aerial installation has to be very large. Briefly, in order to radiate energy efficiently, the length of the transmitting aerial needs to approach at least a quarter of a wavelength at the working frequency.

At a frequency of 10 kHz, the aerial would need to be something approaching 7500 metres (around 4.5 miles), since

$$\lambda = c/f = 3 \times 10^8/10^4 \, \text{m} \qquad \text{i.e. } \lambda/4 = 7500 \, \text{m}$$

Clearly this is not very practical.

So, it is very difficult indeed to transmit low-frequency speech and music information signals directly as radio waves. However, at higher frequencies with shorter wavelengths it becomes easier and more economical to transmit radio waves, so radio systems use high frequencies to "carry" the low-frequency information signals to the destination.

Therefore, in both situations, either for more economic use of bearer bandwidth or to allow radio propagation, the information signals are superimposed on to the carrier signal at the transmitting end by a process called **modulation**. At the destination, the information signal is recovered from the carrier by the reverse process called **demodulation**.

Modulation of a carrier wave is achieved by arranging for some characteristic of the carrier wave to be varied by the information signal. In Chapter 1 it was seen that a sinusoidal a.c. waveform has a number of important characteristics, e.g. peak value (amplitude), frequency, and phase, and it can be arranged for the information signal to vary *any* of these characteristics of the carrier waveform.

In order to simplify the understanding of this modulation process the information signal, or **modulating signal**, will be considered as a single-frequency waveform instead of a band of frequencies as previously considered. The *modulating signal* and the *carrier wave* are applied to a

Fig. 2.1 Simple principle of a modulated telecommunication system

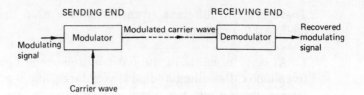

SENDING END   RECEIVING END

Modulating signal → Modulator → Modulated carrier wave → Demodulator → Recovered modulating signal

Carrier wave

modulator circuit, and the **modulated carrier wave** is extracted from the output of the modulator circuit, as illustrated in Fig. 2.1. The function of the demodulator will be considered later.

Amplitude Modulation

This is the process of varying the *amplitude* of the sinusoidal carrier wave by the amplitude of the modulating signal, and is illustrated in Fig. 2.2.

Fig. 2.2 Graphical illustration of an amplitude-modulated carrier wave

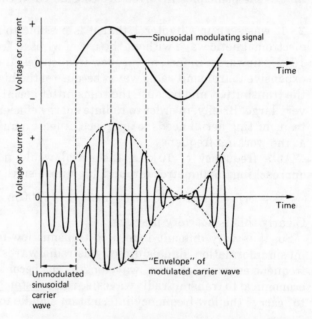

Voltage or current

Sinusoidal modulating signal

Time

Voltage or current

Time

Unmodulated sinusoidal carrier wave

"Envelope" of modulated carrier wave

The unmodulated carrier wave has a constant peak value and a higher frequency than the modulating signal but, when the modulating signal is applied, the peak value of the carrier varies in accordance with the instantaneous value of the modulating signal, and the outline waveshape or "envelope" of the modulated wave's peak values is the same as the original modulating signal waveshape. The modulating signal waveform has been superimposed on the carrier wave.

Frequency Modulation

This is the process of varying the *frequency* of the sinusoidal carrier wave by the amplitude of the modulating signal, and is illustrated in Fig. 2.3.

When the modulating signal is applied, the carrier frequency increases to a maximum value as the modulating signal amplitude increases to a maximum in a positive direction, and decreases to its unmodulated value

Fig. 2.3 Graphical illustration of a frequency-modulated carrier wave

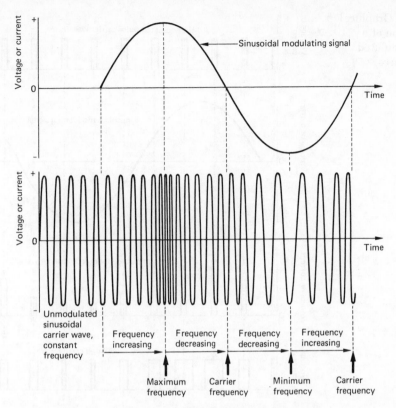

as the amplitude decreases again towards zero. Then, in the second half-cycle of the modulating signal, the carrier frequency decreases to a minimum value as the modulating signal amplitude increases to a maximum in a negative direction, and increases to its unmodulated value as the modulating signal amplitude decreases again towards zero.

Note that the peak value or amplitude of the carrier wave remains constant. It is important to understand that the variation of the carrier frequency above and below its unmodulated value depends on the *amplitude* of the modulating signal voltage (or current).

Pulse Modulation

Another method of conveying information is by means of pulses of voltage or current.

With pulse modulation the carrier wave used is not sinusoidal, but consists of repeated rectangular pulses. The amplitude, width or position of the pulses can be altered by the amplitude of the information signal, as illustrated in Fig. 2.4.

Bandwidth of Amplitude-modulated Carrier Waves

It can be shown by mathematical analysis that, when a sinusoidal carrier wave of frequency f_c Hz is amplitude-modulated by a sinusoidal modulating signal of frequency f_m Hz, then the modulated carrier wave contains *three* frequencies.

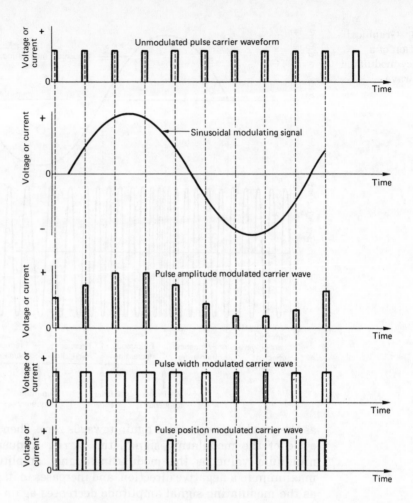

Fig. 2.4 Graphical illustration of a pulse-modulated carrier wave

Unmodulated pulse carrier waveform

Sinusoidal modulating signal

Pulse amplitude modulated carrier wave

Pulse width modulated carrier wave

Pulse position modulated carrier wave

One is the original carrier frequency, f_c Hz.

The second is the *sum* of carrier and modulating signal frequencies:

$$(f_c + f_m)\,\text{Hz}$$

The third is the *difference* between carrier and modulating signal frequencies:

$$(f_c - f_m)\,\text{Hz}$$

This is illustrated in Fig. 2.5.

Fig. 2.5 Simple principle of amplitude modulation

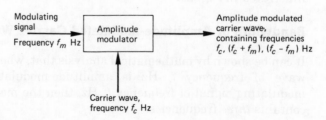

Modulating signal
Frequency f_m Hz

Amplitude modulator

Amplitude modulated carrier wave, containing frequencies f_c, $(f_c + f_m)$, $(f_c - f_m)$ Hz

Carrier wave, frequency f_c Hz

Fig. 2.6 Frequency spectrum of an amplitude-modulated wave for a single-frequency modulating signal

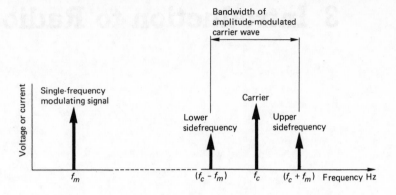

Fig. 2.7 Frequency spectrum of an amplitude-modulated wave for commercial speech modulating signal

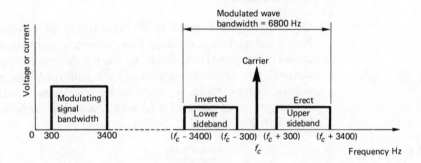

It should be noted that two of these frequencies are new, being produced by the amplitude-modulation process, and are called **sidefrequencies**.

The *sum* of carrier and modulating signal frequencies is called the *upper sidefrequency*.

The *difference* between carrier and modulator signal frequencies is called the *lower sidefrequency*.

This is illustrated in the frequency spectrum diagram of Fig. 2.6.

The *bandwidth* of the modulated carrier wave is

$$(f_c + f_m) - (f_c - f_m) = 2f_m$$

i.e. *double* the modulating signal frequency.

When the modulating signal consists of a *band* of frequencies, as already seen for commercial speech and music for example, then each individual frequency will produce upper and lower sidefrequencies about the unmodulated carrier frequency, and so upper and lower **sidebands** are obtained. This is illustrated in Fig. 2.7.

The bandwidth of the modulated carrier wave is

$$(f_c + 3400) - (f_c - 3400) = 6800\,\text{Hz}$$

which is *double* the *highest* modulating signal frequency.

It follows therefore that, as the modulating signal bandwidth increases, the modulated wave bandwidth also increases, and the transmission system used must be capable of handling this bandwidth throughout.

3 Introduction to Radio Systems

Introduction

One important system used extensively in telecommunication is the *broadcasting* of information from a central point to a wide audience, and called commercial broadcasting. Such a system can operate over lines that connect the central point to all the different destinations or receiving points. For example, a public address system enables information to be broadcast to all parts of a factory or similar large organization, as illustrated in Fig. 3.1.

Fig. 3.1

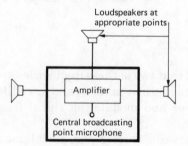

Loudspeakers at appropriate points

Amplifier

Central broadcasting point microphone

This idea of commercial broadcasting can be extended to a much wider audience and over much larger distances by using a radio transmitter to radiate the information through the atmosphere for detection by a receiver anywhere within the range of the transmitted radio signal power. This is well known of course to everyone these days, with commercial radio and television programmes being an inescapable part of life. The general principle of a commercial broadcasting radio system is illustrated in Fig. 3.2.

We have considered radio waves being propagated through the atmosphere without wires. These radio waves can pass through insulating material, although some energy is lost in the process, but are *reflected* by conducting surfaces. These facts can cause the propagation of radio waves to be irregular and unpredictable in performance. Fading of signals due to the presence of large buildings, etc. is a familar effect.

Radio waves can also be propagated through a vacuum, and along a transmission line as already considered. In the case of the line, the energy is transmitted through the insulation separating the conductors.

Fig. 3.2 Elements of an amplitude-modulated sound radio broadcasting system

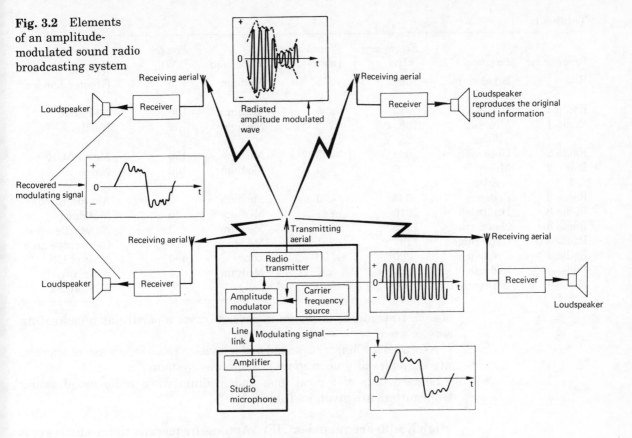

Receiving aerial

Loudspeaker

Receiver

Radiated amplitude modulated wave

Receiving aerial

Receiver

Loudspeaker reproduces the original sound information

Recovered modulating signal

Receiving aerial

Loudspeaker

Receiver

Transmitting aerial

Radio transmitter

Amplitude modulator

Carrier frequency source

Receiving aerial

Receiver

Loudspeaker

Line link

Modulating signal

Amplifier

Studio microphone

Propagation Characteristics of Radio Waves

In Chapter 1 the classification of radio waves was introduced, and each band in the classification had its own particular characteristics.

Low Radio Frequencies (VLF, LF, MF)   At these frequencies the radio energy wave leaves the transmitting aerial and follows the curve of the earth's surface in all directions, usually as a **ground wave**, as illustrated in Fig. 3.3.

Generally speaking, the distance travelled over the earth's surface depends on the *power* generated by the radio transmitter. The power level

Fig. 3.3a Illustration of ground wave propagation over the earth's surface

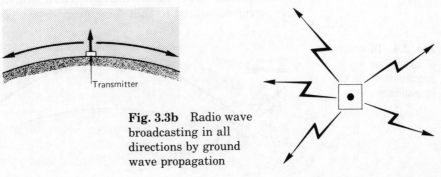

Transmitter

Fig. 3.3b Radio wave broadcasting in all directions by ground wave propagation

Table 3.1

Programme	Location	Frequency (kHz)	Wavelength (metres)	Waveband	Power (kW)	Programme Coverage
Radio 1	Brookmans Park	1089	275	Medium	150	Greater London
Radio 1	Droitwich	1053	285	Medium	150	Midlands
Radio 1	Moorside Edge	1089	275	Medium	150	North
Radio 2	Droitwich	693	433	Medium	150	Most of UK
Radio 2	Moorside Edge	909	330	Medium	100	North
Radio 2	Redmoss	693	433	Medium	2	Aberdeen area
Radio 3	Droitwich	1215	247	Medium	30	Midlands
Radio 3	Newcastle	1215	247	Medium	2	Newcastle area
Radio 3	Cambridge	1197	251	Medium	0.2	Cambridge area
Radio 4	Droitwich	200	1500	Long	400	Most of UK
Radio 4	Redruth	756	397	Medium	2	S. Cornwall
Radio 4	Westerglen	200	1500	Long	50	S. Scotland

of each transmitter is chosen in order to cover a particular broadcasting service area.

As stated in Chapter 1, sound and television radio broadcasting services are examples of *unidirectional* or one-way systems.

Some details of typical long and medium wave radio broadcasting transmitters are given in Table 3.1.

High Radio Frequencies (HF)   At these frequencies the ground wave is absorbed or attenuated very rapidly, but radiation also occurs upwards until the waves reach the **ionosphere**, which extends approximately from 50 to 400 km above the earths surface. In the ionosphere, the gases present are subject to ultraviolet radiation from the sun. The molecules lose some electrons and so become positively charged ions. At certain heights in the ionosphere, recombination between free electrons and positive ions is less likely than in the lower atmosphere, and regions of high ionization exist in the upper atmosphere. Here in these ionized layers, the radio waves are refracted at particular angles in such a way that they are returned to the earth at some distance from the transmitting aerial. This type of radio wave is called a **sky-wave**, and is illustrated in Fig. 3.4.

By using directional aerials at the transmitter, the sky-wave can be made to reach a particular destination at a long distance using relatively low power compared to that for a ground wave over the same distance.

Fig. 3.4 Illustration of sky-wave propagation via the ionosphere

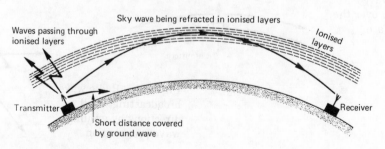

Table 3.2

Station	Radio 1/2 (MHz)	Radio 3 (MHz)	Radio 4 (MHz)	Power (each programme)	Area
Sutton Coldfield	88.3s	90.5s	92.7s	120 kW	Midlands
Ventnor	89.4s	91.6s	93.8	20 W	Isle of Wight (south coast)
Wrotham	89.1s	91.3s	93.5s	120 kW	S.E. England
Morecambe Bay	90.0s	92.2s	94.4s	4 kW	N.W. Lancashire

s – stereo transmission

This method was widely used in the international telephone network for point-to-point communication before long-distance ocean cables and artificial space satellites were introduced.

Very High Frequencies (VHF) At these frequencies the radio wave energy is propagated through space in straight lines, as is light energy. Using **omni-directional aerials** (radiating equally in all directions), BBC vhf/fm sound broadcasting services cover clearly defined areas for Radio 1/2, 3 and 4 programmes, as indicated in Table 3.2.

By using a **directional aerial**, e.g. a parabolic reflector or dish, the energy can be directed towards the horizon, giving a line-of-sight propagation path. This is illustrated in Fig. 3.5.

Fig. 3.5 Illustration of line-of-sight propagation by space wave

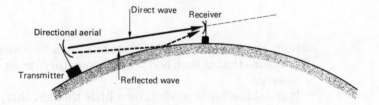

It should be pointed out that space waves can also be directed down to the ground, where they are reflected, and can thus reach the receiving aerial. This *reflected* wave can cause interference to the *direct* wave because of the longer distance it travels, which means that the reflected wave arrives later than the direct wave.

It should also be fairly clear that the distance over which space waves can be used depends on the heights of the transmitting and receiving aerials. So wherever possible these are mounted on masts or towers erected on high ground.

Space-wave propagation is used in the public telephone network for multi-channel microwave radio relay systems as an alternative to the multi-channel coaxial cable links previously mentioned in Chapter 2.

The point-to-point radio systems using sky-waves or space waves can be included in a national or international telephone network, but clearly these systems must be capable of transmitting information in *both* directions if a telephone conversation is to be possible. It was shown in Chapter 1 that in order for this to occur there must be a complete telecommuni-

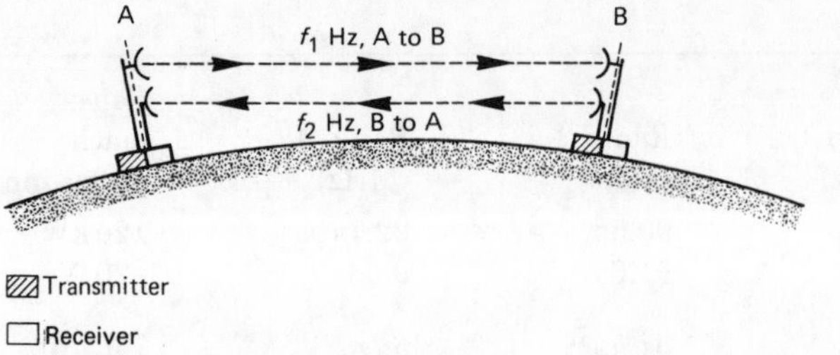

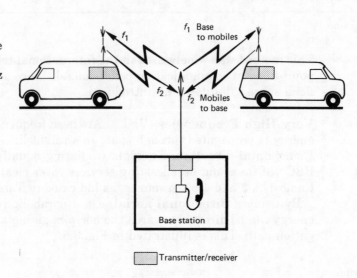

Fig. 3.6 Two-way radio telephone link using line-of-sight space wave (Typical frequencies in the 2 GHz range. Transmitter power typically 10 W.)

A f_1 Hz, A to B   B

f_2 Hz, B to A

▨ Transmitter

☐ Receiver

Fig. 3.7 Two-way mobile radio telephone system
(f_1 typically 164.5 MHz at 15W
f_2 typically 160.0 MHz at 5 W)

f_1

f_1 Base to mobiles

f_2

f_2 Mobiles to base

Base station

▨ Transmitter/receiver

cations channel in both directions to form a two-way or bothway circuit. This means that at each end of the system there must be a transmitter and a receiver.

It should be fairly obvious by a little thought that, if a large amount of transmitter power is needed to enable a radio wave to reach the distant receiver, then the receiver at each end would be subjected to a very much more powerful signal from its own transmitter than that received from the distant end. It is usual, therefore, to allocate different radio carrier frequencies for the two directions of information transmission. This is illustrated in Fig. 3.6. and Fig. 3.7.

Furthermore, with hf point-to-point links over long distances, very large transmitter powers may be required (e.g. 30 kW). In this situation it is usual to separate the transmitter and receiver at each end geographically in order to prevent a receiver being swamped by the transmitter at the same end of the system. This is illustrated in Fig. 3.8.

Communication Satellites

Once it had been clearly demonstrated that artificial satellites could be successfully launched into orbit around the earth, thoughts naturally turned to the possibility of using orbiting satellites for worldwide telephone communication. Originally, long-distance international telephone

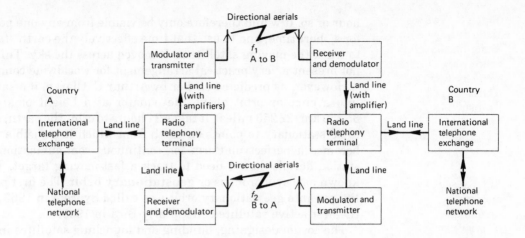

Directional aerials

f₁ A to B

Modulator and transmitter

Receiver and demodulator

Land line (with amplifiers)

Land line (with amplifier)

Country A

Country B

International telephone exchange

Land line

Radio telephony terminal

Radio telephony terminal

Land line

International telephone exchange

Land line

Land line

National telephone network

Receiver and demodulator

Directional aerials

f₂ B to A

Modulator and transmitter

National telephone network

Fig. 3.8
International telephone call routed over two-way high-frequency radio link (Frequencies in the range 3 to 30 MHz. Typical power 30 kW.)

calls depended on hf sky-wave links described on page 20. These were eventually replaced by the introduction of oceanic telephone cables in the 1950s.

In fact, some very early experiments were carried out to bounce signals from the moon, followed by more serious experiments using inflated balloons. These experiments involved the use of a passive body simply to reflect signals back to earth, so that the signals arriving back on earth after the two-way journey were very weak. The first inflated balloon project was called ECHO, launched in 1960.

Clearly there was an obvious advantage in using an orbiting artificial satellite in that it could contain solar-powered equipment to receive the signal transmitted upwards from earth and re-transmit it back to earth from a transmitter located in the satellite. The satellite would therefore be active, and much stronger signals would arrive back on earth than from a purely passive device. By using frequencies in the shf range (3–30 GHz), highly directional dish aerials could be employed both at the earth station and on the orbiting satellites.

It is possible to put a satellite into a number of different orbits around the earth. The orbital path can be circular or elliptical at different heights above the earth, and at different angles, as illustrated in Fig. 3.9.

Early artificial satellites were put into elliptical orbits of various paths just a few hundred miles above the earth, with each orbit taking maybe an

Fig. 3.9 Typical satellite elliptical orbits

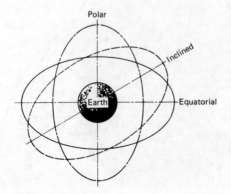

Polar

Inclined

Earth

Equatorial

hour or so. It would therefore only be visible from any one point on earth for a short time, and to use that time effectively the earth station needed to track the moving satellite as it moved across the sky. This clearly did not present a very practical arrangement for worldwide communication.

However, as predicted earlier by Arthur C. Clarke, if a satellite is put into a circular orbit above the equator at a height of approximately 35 750 km (22 250 miles), it takes 24 hours to orbit the earth, so it appears to be stationary to points on earth from which it is visible. It therefore becomes a perfect platform for continuous communication, and earth station aerials do not need to track a fast-moving target. This became known as a synchronous or **geostationary orbit**. The first passive satellite put into geostationary orbit was called Syncom in 1963, followed by the first active satellite called Early Bird in 1965.

The cost of designing, building and launching satellites into orbit was obviously going to be very high. Although some countries did proceed alone originally, eventually many countries set up the International Telecommunication Satellite Consortium (Intelsat) to seriously pursue the development of satellite communication. The original Early Bird satellite was taken over in 1966 and redesignated Intelsat I.

Rapid progress was made in satellite technology, with increase in size, power availability, number of telephone channels, and so on, resulting in Intelsat IV, IVA and V models in use by the early 1980s, with Intelsat VI due to go into service very soon. The satellites themselves are cylindrical in shape, the later ones having extending arms for the solar cells. Intelsat I was about half a metre long, whereas Intelsat VI will be about 22 metres long.

Fig. 3.10 Simple principle of telephone communication by satellite

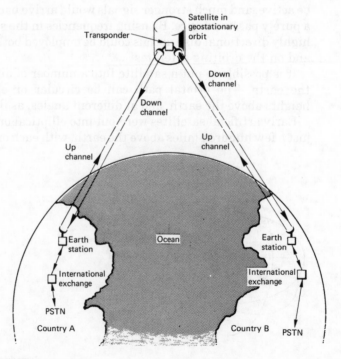

By putting satellites into geostationary equatorial orbits above the Atlantic, Pacific and Indian oceans, it has been possible to set up a rapidly expanding worldwide communication network for telephone and television transmission. By 1982 there were over 100 member countries of Intelsat, with nearly 40 satellites having been launched and over 200 earth stations established. The earth stations are owned by the various controlling administrations in each country, and the satellites are owned by Intelsat. Fig. 3.10 gives a simple illustration of a satellite communication system, and Figs. 3.11, 3.12 and 3.13 show the earth stations served by the Atlantic, Indian and Pacific Ocean satellites respectively.

Fig. 3.11 Atlantic ocean satellite coverage

Because of the very long distances involved and the limited power availability of the satellites, signals arriving at earth stations are very weak, so low-noise sensitive receivers are needed, usually working at low temperatures to reduce noise to an acceptable level.

Operating frequencies are in the bands 4–6 GHz and 11–14 GHz, and for each link between earth station and satellite different frequencies are used for the "up" and "down" channels.

Fig. 3.12 Indian ocean satellite coverage

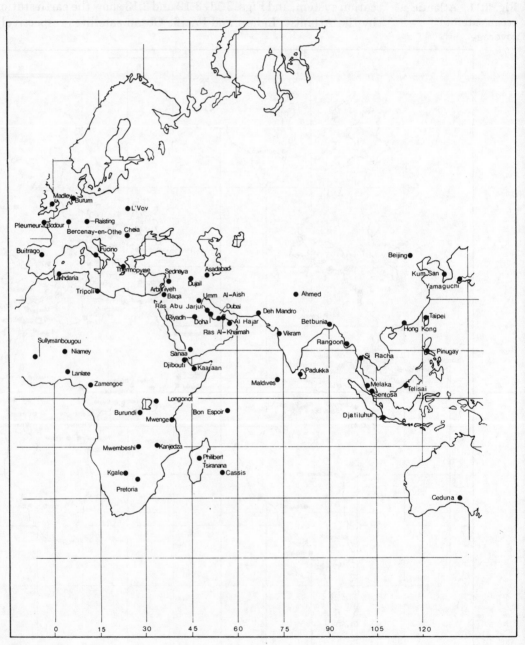

Fig. 3.13 Pacific ocean satellite coverage

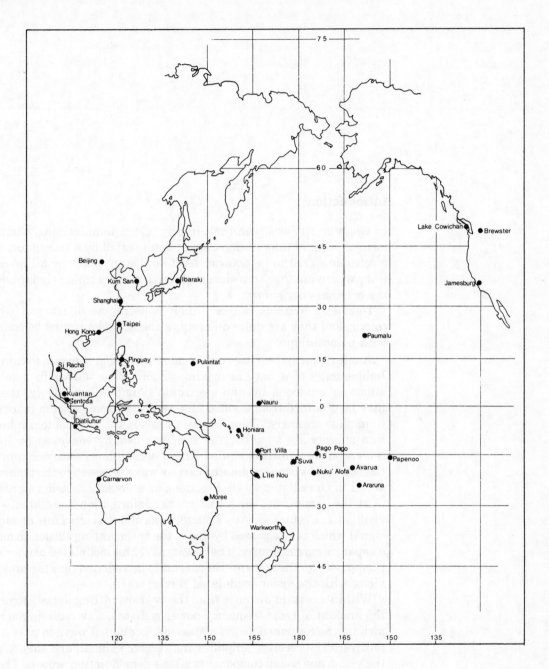

4 Introduction to Television

Introduction

In Chapter 1 it was seen that, in any telecommunication system, some form of original information energy is converted by a transducer into an electronic signal to be transmitted to a distant point by a line or radio link, where another transducer converts the electronic signal back into the original energy form.

Television systems can be either monochrome (black and white) or colour, and they are quite different, although colour must be compatible with monochrome.

A television system uses one or more television cameras to convert the light energy of a natural moving visible scene, either in a television studio or outdoors, into an electronic signal. Alternatively, the signal may be obtained from a video tape recorder, from telecine machines, or from slide scanners. These last two convert films or photographic slides into appropriate signals. This signal is usually conveyed by line to a television transmitting station where it modulates a carrier source, and the resultant vision-modulated carrier wave is passed to the transmitting aerial to be radiated in all directions as a broadcast vision signal.

At the same time, the sound energy information associated with the visible scene is picked up by a microphone and converted into an electronic signal which is also passed by line to the transmitting station to modulate a separate carrier source. The resultant sound-modulated carrier wave is then passed to the transmitting aerial to be radiated into the atmosphere along with the vision-modulated carrier wave.

Within a certain distance from the tv transmitting aerial, according to the amount of radio-frequency power radiated, a tv receiving aerial can pick up the combined vision and sound modulated wave to pass it to a tv receiver. The receiver amplifies the received signal, and then separates the vision and sound components after a demodulation process. The demodulated vision signal is passed to a cathode ray tube to reproduce as closely as possible the original visible moving scene at the transmitting end. The demodulated sound signal is passed to a loudspeaker to reproduce as closely as possible the original sound information associated with the visible scene.

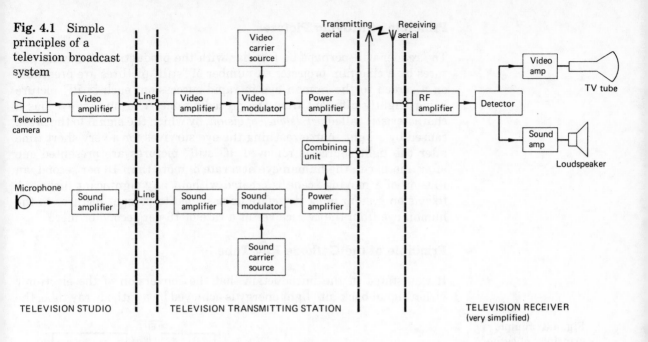

Fig. 4.1 Simple principles of a television broadcast system

Television camera

Video amplifier

Line

Video amplifier

Video modulator

Video carrier source

Power amplifier

Combining unit

Transmitting aerial

Receiving aerial

RF amplifier

Detector

Video amp

TV tube

Sound amp

Loudspeaker

Microphone

Sound amplifier

Line

Sound amplifier

Sound modulator

Sound carrier source

Power amplifier

TELEVISION STUDIO

TELEVISION TRANSMITTING STATION

TELEVISION RECEIVER (very simplified)

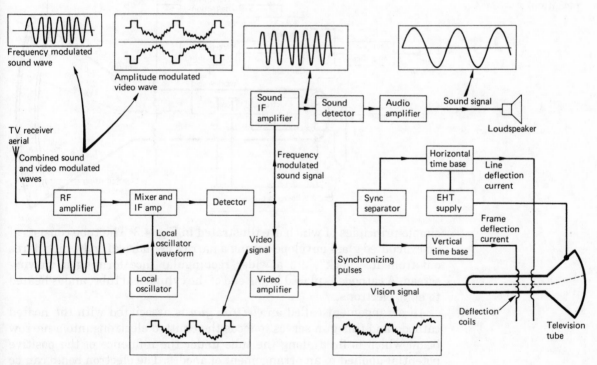

Frequency modulated sound wave

Amplitude modulated video wave

Sound signal

Loudspeaker

TV receiver aerial

Combined sound and video modulated waves

RF amplifier

Mixer and IF amp

Detector

Local oscillator waveform

Local oscillator

Video signal

Video amplifier

Sound IF amplifier

Sound detector

Audio amplifier

Frequency modulated sound signal

Sync separator

Horizontal time base

EHT supply

Vertical time base

Line deflection current

Frame deflection current

Synchronizing pulses

Vision signal

Deflection coils

Television tube

Fig. 4.2 Principles of a television receiver (using f.m. sound signal and negative-modulation vision signal)

The simple principles of a television broadcast system are illustrated in Fig. 4.1 and the basic principles of the tv receiver arrangement are shown in Fig. 4.2.

Principle of Moving Pictures

The reader will perhaps be familiar with the production of moving pictures by a cine-film projector. A number of "still" pictures are presented on a screen to the human eye in rapid succession, each "still" picture being slightly different from the preceding one. The human eye has a characteristic called *persistence of vision*, by which the signal to the brain caused by a light source reaching the eye survives for a very short time after the light source is removed. If "still" pictures are presented one after another to the human eye at a rate of more than 16 per second, an illusion of a moving scene is created without any significant flicker. A television system must therefore be designed to present pictures to the human eye from the tv receiver at a rate of 16 per second or more.

Principle of the Cathode Ray Tube

It was stated in the introduction that the conversion of the electronic vision signal back into light energy is achieved by a cathode ray tube, the

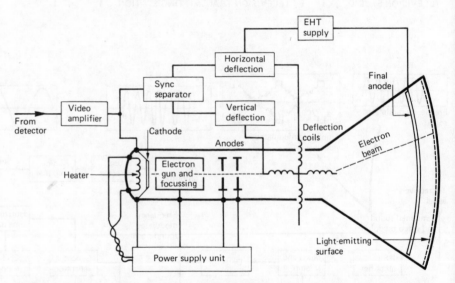

Fig. 4.3 Simple principles of cathode ray tube for television reception

simple principles of which are illustrated in Fig. 4.3. The tube consists of an evacuated glass envelope having a narrow cylindrical end which flares out from its "neck" into a wider rectangular face forming a viewing screen. A cathode is placed in the end of the cylindrical tube, and is heated to emit electrons.

An arrangement called an *electron gun* is associated with the heated cathode and this gun serves to focus the emitted electrons into a narrow beam which is fired along the tube under the influence of the positive potential applied to an arrangement of anodes. The electron beam can be moved in horizontal and vertical directions by magnetic fields produced by currents passing through deflection coils clamped around the outside of the neck of the tube.

The inside surface of the rectangular viewing screen is coated with a light-emitting material. If the electron beam fired along the tube strikes the screen coating with sufficient velocity, the energy of the electron beam causes light to be emitted from the surface coating, and a small spot of light is seen on the tube screen when viewed from the front.

Principle of Scanning

By passing suitable electric currents through the deflection coils, magnetic fields are produced which can control the path of the electron beam along the tube by simultaneous horizontal and vertical forces, and so the small spot of light can be moved around the screen at will. (See Fig. 4.4.)

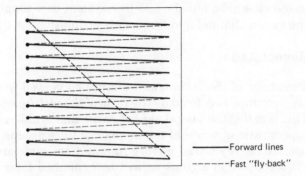

Fig. 4.4 Simple principle of scanning a picture

———— Forward lines

- - - - Fast "fly-back"

To produce a picture, the small spot of light is positioned initially in the top left-hand corner of the rectangular screen as viewed from the front. It is then moved rapidly across the screen by the horizontal deflection force. When the end of the first line is reached, the spot is returned very rapidly to the left-hand side of the screen but positioned slightly below the starting point of the first line. This very rapid return is called the **fly-back** of the spot of light. A second line is now traced out by the horizontal deflection force, and again the fly-back is carried out. The positioning of the spot at the beginning of each line just below the previous line is achieved by the vertical deflecting force. This process is repeated until the spot of light reaches the bottom right-hand corner of the rectangular screen, and one complete picture has been traced out or scanned by the spot of light in successive horizontal lines.

The spot is now returned to the top left-hand corner of the screen, and a second picture is traced out or scanned in the same way as the first one. If this is repeated rapidly so that more than 16 pictures per second are traced out, and if the intensity of the spot of light is constant, the moving spot appears as a complete white picture or *raster*.

The vision signal, demodulated from the received television signal, is now applied to the cathode ray tube to control the intensity of the electron beam passing along the tube. The amount of light energy emitted by the screen material will vary in accordance with the intensity of the electron beam and so the light energy output from the tv screen will be a reproduction of the light energy picked up by the tv camera or other equipment, and the illusion of a moving scene is presented to the observer. This picture appears in black, white and all intermediate shades of grey, and is called a monochrome picture.

Number of Lines

The number of lines used in television systems has varied in different countries over the years. For example, 405, 525, 625 and 819 lines have been used.

In the UK the original BBC and ITA channels in the vhf Band I and Band III (ranging approximately from 30–300 MHz) used 405 lines, but tv transmitting stations now use 625 lines for BBC and ITA channels in the uhf Band IV and Band V (ranging approximately from 300–3000 MHz).

The 405-line system uses amplitude-modulation for both vision and sound channels, but the 625-line system uses amplitude-modulation for the vision channel and frequency-modulation for the sound channel.

Aspect Ratio

The shape of the rectangular picture as viewed from the front of the tv receiver tube is defined by the ratio of the picture width to picture height. This is called the **visual aspect ratio**, and is different from the electrical aspect ratio in terms of the number of lines because not all the lines are used to convey actual picture information; some are needed to transmit synchronization signals as will be explained later in this chapter. The visual aspect ratio used in both systems in the UK is 4:3.

Interlaced Scanning

In addition to at least 16 pictures per second being necessary to create the illusion of a moving picture, it has also been found that the number of pictures per second must be the same as the frequency of the a.c. mains supply in order to avoid "hum bars" appearing across the screen. So in the UK and Europe, 50 pictures per second are needed, and in the USA 60 pictures per second.

Using this number of pictures per second with the simple principle of scanning results in a vision electronic signal having a very large frequency bandwidth, and this would mean that a limited number of tv transmitting stations could be accommodated in the allocated frequency bands. So it may not be possible to provide complete coverage of a particular country.

In order to reduce the bandwidth of the vision signal and therefore allow more tv transmitters to be used, a technique called **interlaced scanning** has been devised. Each complete picture is divided into two frames or fields which are scanned and transmitted one after the other, and then reassembled at the tv receiver. So although 50 frames or fields are transmitted every second to avoid mains "hum bars" (in the UK), only 25 complete pictures are transmitted every second. The vision signal therefore contains only half the information compared with 50 pictures per second, and so the bandwidth is also reduced by a half. This allows twice as many tv transmitters to be allocated in the available bands.

The interlaced scanning of the two frames or fields making up each complete picture is achieved by scanning *alternate* lines by the spot on the

Fig. 4.5 Principle of interlaced scanning of alternate lines

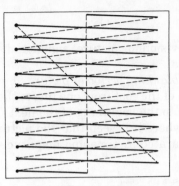

tv tube screen. After one field of alternate lines has been scanned, the spot returns to fill in the gaps between these lines as it scans the second field. This principle is illustrated simply in Fig. 4.5.

Producing the Vision Signal

We have seen that the variations of the vision signal level or strength control the intensity of the electron beam in the tv receiver tube to produce the appropriate amount of light from the tv screen. At the other end of the tv system, the tv camera produces this vision signal by using the same interlaced scanning principle as described for the tv tube.

Very simply, the visible scene to be transmitted is focused by the optical lens system of the tv camera on to a light-sensitive surface which absorbs light energy according to the instantaneous scene being focused by the tv camera.

An electron beam scans the light-sensitive surface and the strength of the electron beam is varied by the amount of light energy absorbed by each small spot on the light-sensitive surface. This variation of the electron beam strength is converted into a varying voltage that constitutes the vision signal containing a range of frequencies which is much wider than that for speech or music. As a guide it can be stated that the *bandwidths* of vision signals in the UK systems are as follows:

For the 405-line system   0 to 3 MHz
For the 625-line system   0 to 5.5 MHz

Synchronizing the TV Camera and the TV Receiver Tube

From the simple principle of interlaced scanning it should be fairly obvious that the electron beam scanning the screen of the tv receiver tube must be in exactly the same position at all times as the electron beam that is scanning in the tv camera. This is achieved by including **synchronizing pulses** along with the vision information signal. Both **line** and **frame** (for field) synchronizing pulses are used to ensure that the correct line of the appropriate frame is being produced by the tv receiver tube. These synchronizing pulses are separated from the vision signal in the tv receiver, and are used to trigger line and frame time-base circuits which supply the currents for the deflection coils to position the spot of light on the tv receiver screen. This is illustrated in Fig. 4.2 given earlier.

The Video Signal

The combination of vision (or picture) signal and synchronizing pulses is called a **video signal**.

Also included is a *blanking period* or *picture suppression period*, to allow time for the fly-back of the spot from one line to the next and from the end of one field or frame to the beginning of the next. This is illustrated very simply in Fig. 4.6, where it is seen that the picture and synchronization information are separated by time and amplitude.

Fig. 4.6 Simple principle of video signal

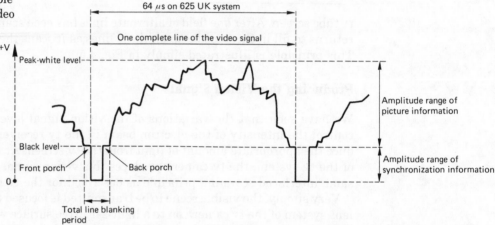

Fig. 4.6 illustrates what is known as a *positive-going* (or positive modulation) video signal. It should be pointed out that some tv systems use a *negative-going* (or negative modulation) video signal which is upside-down compared with Fig. 4.6.

The line and field (or frame) synchronization pulses are included in the blanking level amplitude region. *Line* synchronization pulses are simple narrow pulses, whilst field synchronization pulses are a series of broader pulses. This is illustrated simply in Fig. 4.7.

The field synchronization pulses take a time equivalent to a number of lines, according to the system being used, and are followed by a number of suppressed lines.

In the *UK 405-line system*, the field sync pulses occupy the equivalent of 4 lines, followed by 10 suppressed lines, giving an overall suppression

Fig. 4.7 Simple illustration of line and field sync-pulses

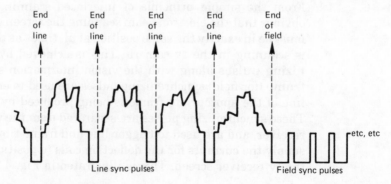

equivalent to 14 lines per field. So, for each complete picture consisting of 2 successive interlaced fields, 28 lines are used for field synchronization giving (405 − 28) = 377 lines containing the picture or vision information.

In the *UK 625-line monochrome system*, 20 lines are used for each field synchronization, giving (625 − 40) = 585 lines containing the picture or vision information .

Bandwidth of Video-modulated Carrier Signals

It was seen in Chapter 2 that when a carrier is amplitude-modulated by an information signal, the bandwidth of the modulated wave is *double* the highest modulating signal frequency. It was also stated, earlier in this chapter, that a vision signal has a wide range of frequencies, with a highest frequency of approximately 3 MHz for a 405-line system and 5.5 MHz for a 625-line system. Using normal double-sideband amplitude modulation would therefore require a bandwidth of approx 6 MHz for 405 lines and 11 MHz for 625 lines, *plus* a bandwidth for the separate sound-modulated carrier waveform.

To reduce the bandwidth needed for each transmission and therefore allow more transmissions in a given frequency band, the modulated wave is passed through a **vestigial sideband filter**, which suppresses part of *one* of the sidebands. This is illustrated simply in Fig. 4.8.

In Fig. 4.8, frequencies are shown relative to the allocated vision carrier frequency. It will be seen that in a 405-line system the sound carrier is 3.5 MHz *below* the associated vision carrier, and the *upper* sideband is restricted. By comparison, in a 625-line system, the sound carrier is 6 MHz *above* the associated vision carrier, and the *lower* sideband is restricted. Notice also in Fig. 4.8 that there is a "guard edge" between the sound information bandwidth and the extreme edge of the particular channel.

In the UK, vision and sound carrier frequencies are allocated to the various channels in the tv bands in such a way that each complete channel slots into a particular section of the band so that interference is avoided between adjacent channels. Typical tv channel frequencies are given in Table 4.1.

Table 4.1

Band 1 (405-lines) *(40–68 MHz)*			*Band IV (625-lines)* *(470–610 MHz)*			*Band V (625-lines)* *(610–940 MHz)*		
Channel	*Carrier Frequencies (MHz)*		*Channel*	*Carrier Frequencies (MHz)*		*Channel*	*Carrier Frequencies (MHz)*	
	Sound	*Vision*		*Sound*	*Vision*		*Sound*	*Vision*
1*	41.5	45.0	21	477.25	471.25	39	621.25	615.25
2	48.25	51.75	22	485.25	479.25	40	629.25	623.25
3	53.25	56.75	23	493.25	487.25	41	637.25	631.25
4	58.25	61.75	24	501.25	495.25	42	645.25	639.25
5	63.25	66.75	25	509.25	503.25	43	653.25	647.25
*originally double sideband			34	581.25	575.25	68	853.25	847.25

Fig. 4.8 Bandwidths of typical TV systems in UK

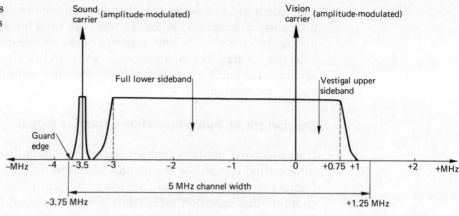

Sound carrier (amplitude-modulated)

Vision carrier (amplitude-modulated)

Full lower sideband

Vestigal upper sideband

Guard edge

-MHz   -4   -3.5   -3   -2   -1   0   +0.75 +1   +2   +MHz

-3.75 MHz

5 MHz channel width

+1.25 MHz

(a)   405-LINE SYSTEM

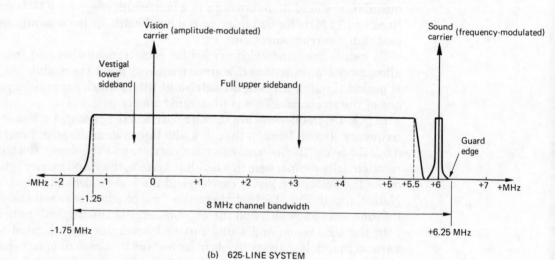

Vision carrier (amplitude-modulated)

Sound carrier (frequency-modulated)

Vestigal lower sideband

Full upper sideband

Guard edge

-MHz -2   -1   0   +1   +2   +3   +4   +5 +5.5 +6   +7   +MHz

-1.25

-1.75 MHz

8 MHz channel bandwidth

+6.25 MHz

(b)   625-LINE SYSTEM

Introduction to Colour Television

When a colour television service is introduced in any country, it is clearly uneconomical to consider providing separate colour and monochrome services. Some viewers will obviously prefer to retain their monochrome receivers, if only because of cost. So it is necessary to provide a service that allows colour or monochrome viewing by choice of domestic receiver. This requires a signal from the television camera that can be received by either a colour or monochrome receiver, as required. Such a signal is called a **compatible signal**, and is produced by the television camera in *two* distinct parts:

the **luminance** part
the **chrominance** part

The *luminance* part contains the BRIGHTNESS information similar to that already described for a monochrome-only system. The *chrominance* part contains the ADDITIONAL information needed for a colour system.

So, monochrome receivers use only the luminance part of the compatible signal, but colour receivers use both luminance and chrominance parts of the compatible signal.

There are three ways in which the compatible signal can be produced from a tv camera to provide the luminance and chrominance signals:

1   The NTSC system (National Television Systems Committee), developed and introduced in the United States of America in the early 1950s, and later also adopted in Canada, Japan and Mexico.
2   The PAL system (Phase Alternate Line), developed in West Germany from the NTSC system, and later also adopted in the United Kingdom and other European countries.
3   The SECAM system (Sequential Colour And Memory), developed in France, and later also adopted in East Germany, USSR and other countries in Europe and North Africa.

In all three systems, the luminance and chrominance information signals from the tv camera are combined to form a compatible video signal, including line and frame synchronizing pulses, which then modulates a vision carrier frequency for transmission on a particular tv channel. The resulting modulated wave has to be accommodated in the same radio frequency bandwidth allocated for any 625-line monochrome television channel as described earlier.

The chrominance information signal is combined or interleaved with the luminance information signal at the transmitter by a process called **encoding**, and at the receiver the reverse process called **decoding** is used.

This chapter will deal only with the principles of the PAL system in service in the United Kingdom.

Properties of Colours

Every colour has three important properties: luminance (brightness), hue (colour) and saturation (amount of colour).

It will be remembered that the brightness or luminance information in a monochrome system is produced by a *single* electron tube in the tv camera. In a colour tv system, the camera contains at least *three* tubes, one each for the colours red, green and blue. These are the predominant or **primary colours** of the rainbow, produced when direct sunlight (white light) is passed through a cloud of moisture (or a prism) and is dispersed into these three primary colours and other intermediate colours (see Fig. 1.11, page 9).

By a reverse process, the colours of the rainbow can be recombined to form white light.

Complementary Colours

When red and green light are mixed, yellow is seen by the human eye. When green and blue are mixed, cyan is seen. When red and blue are mixed, magenta is seen.

The yellow, cyan and magenta are called *complementary colours*.

By combining red, green and blue light in various intensities, it is possible to produce for the human eye any of a wide range of natural colours. The three primary colours of red, green and blue are extracted from the natural scene to be televised by using suitable light filters in front of each of the three camera tubes.

The luminance and chrominance signals are then produced from the three primary colour signals in the *encoding* circuits, and combined to form the compatible video signal which is then passed to the modulator of the tv transmitter.

At a colour receiver, after demodulation, the compatible signal is applied from the *decoding* circuits to a three-gun cathode ray tube to reproduce the colour picture.

At a monochrome receiver, after demodulation, the luminance part of the compatible signal is applied to a single-gun crt to reproduce a black and white picture as already described.

The simple arrangement of a colour tv system is illustrated in Fig. 4.9.

Fig. 4.9 Simple principles of a colour TV broadcast system

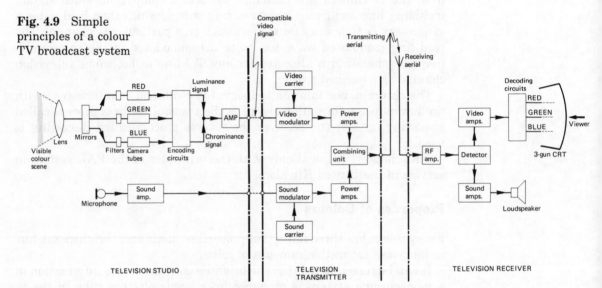

The Luminance Signal

This contains the *brightness* information, produced by combining in the encoder the red, green and blue signals, from the tv camera tubes and filters, in the particular proportions required by the human eye to observe white light.

This proportion is 30% red, 59% green and 11% blue.

The luminance signal is generally designated by Y, and can therefore be represented by the equation

$$Y = 0.3R + 0.59G + 0.11B$$

If such a signal is applied to a monochrome receiver single-gun crt, the light output produces a black and white picture that is virtually identical to that produced originally from a single-tube monochrome tv camera.

The Chrominance Signal

As stated previously, this *additional* information for the colour signal is placed into the normal bandwidth of a 625-line channel that is produced by modulating the allocated channel vision carrier frequency by the luminance signal.

The chrominance signal modulates a *sub-carrier frequency*, and this sub-carrier is suppressed at the transmitter. The sub-carrier frequency is approximately 4.43 MHz, and the bandwidth produced by the chrominance signal modulation is approximately 1 MHz above and below the sub-carrier frequency. This will be illustrated later.

The chrominance signal is obtained in a particular way by using **colour-difference signals**, which are

Red minus luminance $(R - Y)$
Green minus luminance $(G - Y)$
Blue minus luminance $(B - Y)$

By modulating the sub-carrier frequency with any *two* of these colour-difference signals, the third can be extracted by the receiver decoder. The $(R - Y)$ and $(B - Y)$ signals are used to modulate the sub-carrier frequency, and the $(G - Y)$ signal is extracted at the receiver.

The $(R - Y)$ signal is usually called the *V-component* of the chrominance signal, and the $(B - Y)$ signal is called the *U-component* of the chrominance signal, with particular control or *weighting* of the signal levels being applied before they modulate the sub-carrier frequency.

$$V = 0.877 \ (R - Y) \quad \text{and} \quad U = 0.493 \ (B - Y)$$

The sub-carrier frequency is applied directly from the generator to the U-modulator, but it is applied to the V-modulator through a 90° phase-shift network followed by a phase-reversing network. This network reverses the phase of the sub-carrier for the duration of *alternate* lines of the frame scan. This process is used to provide automatic correction of phase distortion that would otherwise result in incorrect hues or colours being reproduced.

In order to synchronize the receiver decoder to the suppressed sub-carrier for demodulation of the U and V signals, it is necessary to transmit a few cycles of the sub-carrier frequency at particular times. These signals are called **colour bursts**, and approximately 10 cycles are transmitted during the "back-porch" period of the line synchronizing pulses of the video signal.

Fig. 4.6 gives a simple illustration of a *positive-going* or *positive-modulation* video signal from a monochrome tv camera. The back-porch period of the line pulse is clearly indicated. Such a signal can be inverted to give a *negative-going* or *negative-modulation* video signal, as mentioned on page 34, and is used in the UK 625-line system.

The simple principle of adding chrominance information signals and sub-carrier colour bursts to a negative-modulation luminance signal is illustrated in Fig. 4.10. The method of achieving this addition is shown in the simple block diagram of Fig. 4.11, and the resulting channel signal bandwidth is illustrated in Fig. 4.12. This can be compared with Fig. 4.8b for a 625-line monochrome channel.

The simplified block diagram of a colour tv receiver, that will respond to the channel signal illustrated in Fig. 4.12, is shown in Fig. 4.13.

Fig. 4.10 Simple illustration of a negative-going compatible video signal

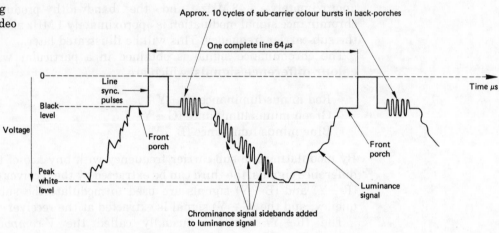

Fig. 4.11 Principle of producing a compatible video signal

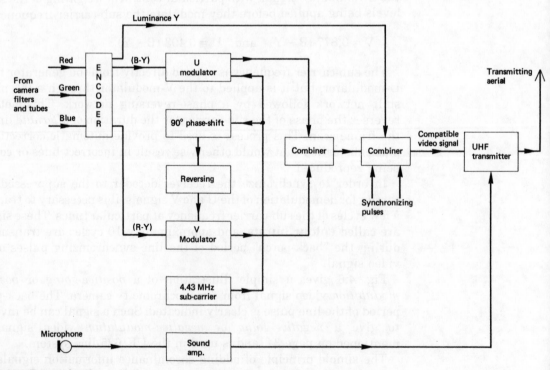

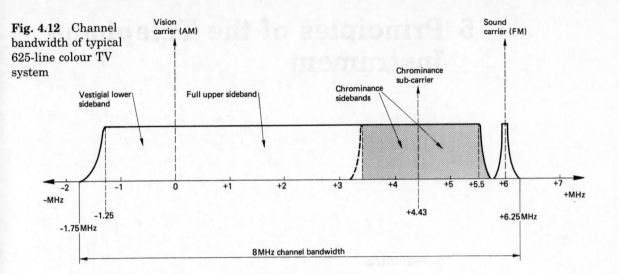

Fig. 4.12 Channel bandwidth of typical 625-line colour TV system

Vision carrier (AM)

Sound carrier (FM)

Chrominance sub-carrier

Chrominance sidebands

Vestigial lower sideband

Full upper sideband

-2 −MHz   -1   0   +1   +2   +3   +4   +5   +5.5   +6   +7 +MHz

-1.25

-1.75MHz

+4.43

+6.25 MHz

8 MHz channel bandwidth

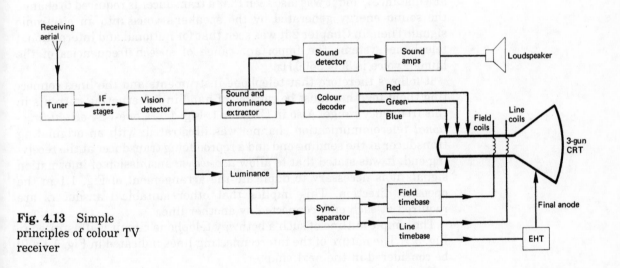

Fig. 4.13 Simple principles of colour TV receiver

Receiving aerial

Tuner → IF stages → Vision detector → Sound and chrominance extractor

Sound detector → Sound amps → Loudspeaker

Colour decoder → Red, Green, Blue

Luminance

Sync. separator → Field timebase → Line timebase → EHT

Field coils, Line coils, 3-gun CRT, Final anode

5 Principles of the Telephone Instrument

Introduction

In Chapter 1, the origin of the word "telephone" was seen to be "speaking at a distance", and it was also seen that a transducer is required to change the sound energy generated by the speaker's voice into an electronic signal. Then, in Chapter 2 it was seen that for national and international telephone systems the important range of speech frequencies of the human voice is 300–3400 Hz.

It follows therefore that telephone instruments and the lines connecting them together must be capable of handling alternating currents in this frequency range. Also in Chapter 1, Fig. 1.1, a one-way or *unidirectional* telecommunication channel was illustrated, with an originating transducer at the sending end and a reproducing transducer at the receiving end. It was stated that to allow *two-way* transmission of information signals it is necessary to duplicate the arrangement of Fig. 1.1 in the opposite direction. This implies that other suitable transducers are required at each end, connected by another line.

The simple outline of such a bothway telephone circuit is illustrated in Fig. 5.1. The nature of the interconnecting lines indicated in Fig. 5.1 will be considered in the next chapter.

For a number of reasons, one of them being economy in the provision of these lines, it was decided to produce a telephone system in which the electronic speech information signals in both directions are carried by a *single* line, as illustrated in Fig. 5.2.

The decision to use this arrangement has a very important influence in the design of a suitable telephone instrument, and in fact was the direct cause of perhaps the most difficult problem encountered, the one of *sidetone*. Briefly, the problem arises from the fact that speech energy signals generated by a sending transducer are passed to the associated receiving transducer as well as being passed to line.

The Sending Transducer

The reader will no doubt be familiar with the use of the **microphone** as a transducer which picks up sound energy waves and converts them into

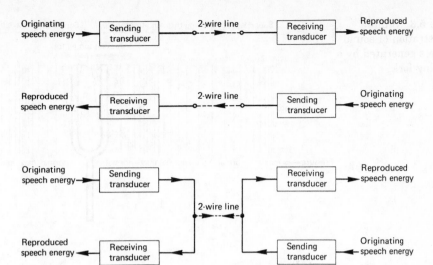

Fig. 5.1 Simple principle of two-way telephone circuit, using two lines

Originating speech energy → [Sending transducer] — 2-wire line — [Receiving transducer] → Reproduced speech energy

Reproduced speech energy ← [Receiving transducer] — 2-wire line — [Sending transducer] ← Originating speech energy

Fig. 5.2 Simple principle of two-way telephone circuit, using a single interconnecting line

Originating speech energy → [Sending transducer] — 2-wire line — [Receiving transducer] → Reproduced speech energy

Reproduced speech energy ← [Receiving transducer] — [Sending transducer] ← Originating speech energy

electronic signals. This word "microphone" is widely used in public address systems, radio and television broadcasting, tape recorders, and so on. But as the telephone system developed, the term **telephone transmitter** came to be accepted for general use.

There are several types of microphone now in use for the systems mentioned previously, but the one which has been used for many years as a standard for the telephone is the **carbon granule** transmitter (although different types of unit are now used in several modern instruments).

The Carbon Granule Transmitter

To understand the simple principles of the transmitter it is necessary to consider the nature of the sound energy waves produced by the human voice. Normally, at ground level, the atmosphere can be considered as columns of air having a normal pressure of approximately 1.05 kilogrammes per square centimetre (15 pounds per square inch). Any source of sound energy has some element which vibrates and causes variations of atmospheric pressure above and below the normal value, and these variations are passed through the atmosphere as sound energy waves gradually decreasing in value until the energy is used up.

One example of a simple source of sound is a tuning fork, which vibrates at a particular audio frequency according to its physical size. The way in which sound energy waves can be represented is illustrated in Fig. 5.3.

When a person speaks, the vocal chords set up vibrations of the columns of air which produce a speech sound information signal. These vibrations reach the telephone transmitter where a *diaphragm* responds to the vibrations and begins to vibrate itself. Increases in pressure move the diaphragm inwards, and decreases in pressure allow the diaphragm to move outwards. The simple principle of this diaphragm is illustrated in Fig. 5.4.

Fig. 5.3 Simple illustration of sound waves generated by a tuning fork

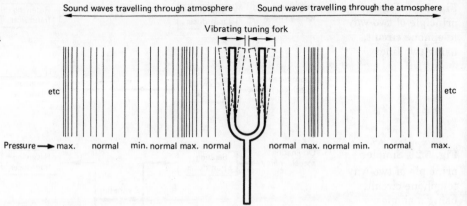

Sound waves travelling through atmosphere   Sound waves travelling through the atmosphere

Vibrating tuning fork

etc   etc

Pressure ➡ max.   normal   min. normal max. normal   normal max. normal min.   normal   max.

Fig. 5.4 Simple principle of telephone transmitter diaphragm

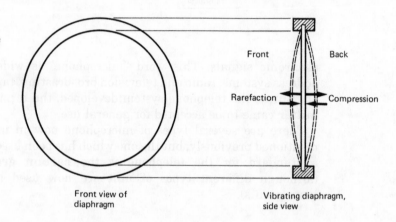

Front   Back

Rarefaction   Compression

Front view of diaphragm

Vibrating diaphragm, side view

The vibrations of the diaphragm are now used to produce a varying electric current that forms an electronic information signal that is ideally the direct copy or analogue of the speech information energy.

It is necessary now to recall briefly the relationship between voltage, current and resistance in an electrical circuit. If a source of electrical energy (e.g. a battery), which has an electromotive force (e.m.f.) of E volts and zero internal resistance, is connected to a circuit which has a resistance (opposition to current flow) of R ohms, then the value of electric current, I amperes, flowing in the circuit is given by

$$\text{Current flowing} = \frac{\text{Electromotive force}}{\text{Resistance}}$$

Using standard symbols this is written as $I = \dfrac{E}{R}$

The circuit diagram for this is given in Fig. 5.5.

Therefore, for a given value of e.m.f., the current will increase if the resistance is reduced, and the current will decrease if the resistance is increased.

Fig. 5.5 Illustration of simple electric circuit

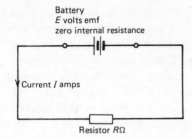

Battery
E volts emf
zero internal resistance

Current I amps

Resistor RΩ

Now, if it can be arranged for the vibrations of the transmitter diaphragm to vary the resistance of an electric circuit, then the current in the circuit will vary in sympathy with the diaphragm as it vibrates due to the speech sound energy waves. This is achieved by attaching a carbon block or electrode to the diaphragm, and placing this electrode inside a chamber containing hard polished carbon granules. Another carbon block or electrode is fixed inside the chamber. The simple principle is illustrated in Fig. 5.6, and the simple equivalent circuit is shown in Fig. 5.7 where the carbon granule chamber and diaphragm are represented by a variable resistance, R ohms.

The form of the varying electric current produced by the diaphragm vibrating under sound energy waves is given in Fig. 5.8.

Fig. 5.6 Simple principle of carbon granule telephone transmitter

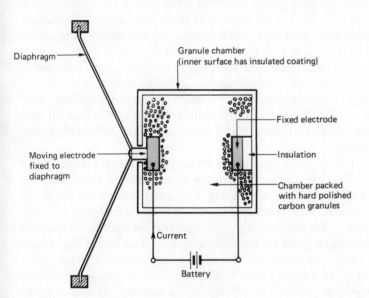

Diaphragm

Granule chamber
(inner surface has insulated coating)

Fixed electrode

Moving electrode fixed to diaphragm

Insulation

Chamber packed with hard polished carbon granules

Current

Battery

Fig. 5.7 Simple electric circuit representing carbon granule telephone transmitter

Variable resistor
RΩ

Current I amps

Battery
E volts

Fig. 5.8 Varying d.c. produced by carbon granule transmitter

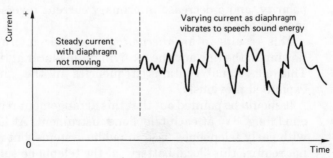

Current

+

Steady current with diaphragm not moving

Varying current as diaphragm vibrates to speech sound energy

0

Time

Fig. 5.9 Illustration of varying resistance of carbon granule telephone transmitter

(a) Normal pressure, normal contact area, normal resistance.

(b) Increased pressure, increased contact area, reduced resistance

(c) Decreased pressure, decreased contact area, increased resistance.

The variation of resistance is due to the fact that the varying pressure on the hard polished carbon granules produces different areas of contact between adjacent granules, as illustrated in Fig. 5.9. The battery is therefore essential in order to provide the direct current flowing through the carbon granule transmitter, otherwise the transmitter cannot function. This direct current is called the *polarizing current*. Some types of transmitter or microphone do *not* require this polarizing current, but the carbon granule transmitter does.

Fig. 5.10 Simple one-way telephone circuit

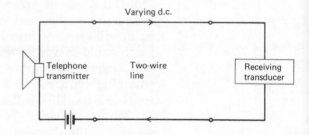

One way of passing the electronic speech information signal to the receiving transducer of the distant telephone is shown in Fig. 5.10. (This diagram uses a pictorial representation of a telephone transmitter, not the standard symbol.) In this arrangement the resistance of the carbon granule transmitter is connected in series with the resistance of the line, which consists of two wires insulated from each other and from earth.

If the line is long, the line resistance may be much greater than the transmitter resistance, and so the *variations* of the transmitter resistance as the diaphragm vibrates will be very small compared with the total circuit resistance. The variations of current will also be very small and the receiving transducer at the distant telephone will not be able to respond satisfactorily.

This difficulty can be overcome by using a battery of much higher e.m.f., or by isolating the transmitter resistance from the line resistance by means of a transformer, as shown in Fig. 5.11 (which uses the standard symbol for a telephone transmitter). The action of the transformer is such that, with a steady current flowing in the primary winding, no current flows in the secondary winding connected to line. But when the transmitter current varies, an e.m.f. is induced into the secondary winding, by mutual inductance, to drive a current in the line. An increase in primary current produces an induced e.m.f. in the secondary with a certain polarity, and a decrease in primary current reverses the polarity of the induced e.m.f.

This results in an *alternating* electronic information signal current flowing in the line and the receiving transducer at the distant transducer. This a.c. signal contains frequencies in the range 300–3400 Hz, as explained previously.

It should be pointed out that this arrangement requires a battery of low e.m.f. (e.g. 3V) at each telephone instrument. At least this was the case with early telephones used in public systems, but modern telephones do not require this "local battery" at the telephone subscriber's premises. It

Fig. 5.11 Use of transformer (induction coil) in simple one-way telephone circuit

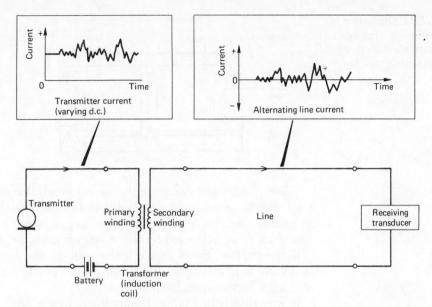

Transmitter current (varying d.c.)

Alternating line current

Transmitter

Primary winding

Secondary winding

Line

Receiving transducer

Battery

Transformer (induction coil)

should also be pointed out that the transformer shown in Fig. 5.11 is, by accepted practice in telephone language, called an "induction coil".

The Receiving Transducer

Several types of transducer have been used over the years for reproducing the speech sound information energy from the alternating electronic speech information signal. The type used in modern telephone instruments is called the **rocking armature receiver**.

Principle of the Rocking Armature Receiver   It consists essentially of a permanent bar magnet with extended soft iron yoke and pole pieces, as shown in Fig. 5.12. Coils of insulated wire are wound around the pole pieces, and these coils are connected in series with the two-wire line from the distant telephone. An armature is pivoted at its centre and arranged so that it is held horizontal by the permanent magnetic field as long as no current is flowing from line into the coils, which is the situation when the magnetic fields in the gaps between the pole pieces and the armature are equal.

Fig. 5.12 Simple principle of rocking armature receiver

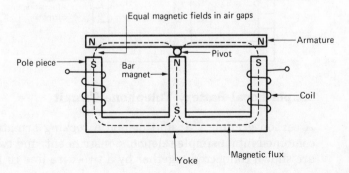

Equal magnetic fields in air gaps

Armature

Pole piece

Pivot

Bar magnet

Coil

Yoke

Magnetic flux

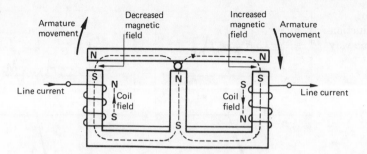

When a current flows from the line through the receiver coils, electro-magnetic fields are produced by the coils such that the field in one gap is increased and the field in the other gap is decreased. This causes the armature to be attracted by the strongest field, as shown in Fig. 5.13.

If the current flows through the coils in the opposite direction, the magnetic field strengths in the gaps are reversed, and the armature is attracted by the opposite pole face.

It was shown in Fig. 5.11 that the action of the induction coil at the sending end produces an alternating electronic speech current that contains frequencies in the range 300–3400 Hz. So the current passing through the receiver coils is constantly reversing direction, causing the armature to rock on its pivot from one pole face to another in sympathy with the alternating speech currents from the line. The movements of the armature are transmitted to a diaphragm by a driving pin, as illustrated in Fig. 5.14, and the diaphragm vibrates to generate a sound energy wave that is a reasonable reproduction of the original sound energy information generated by the person talking at the other end of the line.

Fig. 5.14 Reproduction of sound energy by rocking armature receiver

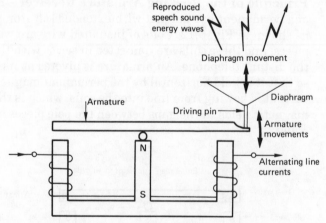

Simple Local-Battery Telephone Circuit

A carbon granule transmitter and a rocking armature receiver can be combined into a simple telephone instrument, and two such instruments are shown connected together by a two-wire line in Fig. 5.15.

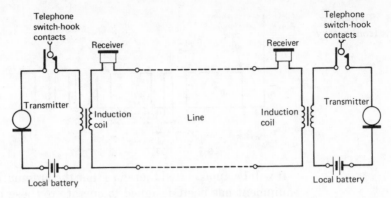

Fig. 5.15 Simple local battery telephone circuit

It will be seen from Fig. 5.15 that if a person is speaking at the left-hand telephone, the alternating speech signal in the line flows through *both* receivers in series, so the talker will hear his own voice. As previously mentioned, this is called sidetone, and has a number of practical disadvantages. In a modern telephone instrument the design is such that the level of sidetone is reduced to an acceptable minimum.

Remember also that the design dispenses with a local battery, but the polarizing current for the carbon granule transmitter is obtained from a central battery located in the telephone exchange to which the telephone instrument is connected.

As well as the carbon granule transmitter (microphone) and rocking-armature receiver used for speaking and hearing, components are needed for other basic functions of the telephone instrument.

Gravity switch   This is operated by removing and replacing the telephone handset on its cradle. Removal of the handset operates electrical contacts that complete a d.c. loop condition to signal to the automatic exchange equipment that a call is being initiated. Replacement of the handset disconnects the loop to signal that the call has been completed.

Dial   This is a rotary-operating mechanical device that sends signals to the already-seized automatic-exchange equipment to indicate the telephone number to which connection is needed. The required number is selected one digit at a time by operating the dial in a clockwise direction, and the appropriate signal for each digit is sent to the exchange switching equipment as the dial restores in an anticlockwise direction to its normal position under the control of a return spring. The dialled signals consist of a number of disconnections of the d.c. loop condition to represent the selected digit. Loop-disconnect pulses are generated at the rate of 10 per second with a 2-to-1 break-to-make ratio. So each pulse consists of a "break" of $66\frac{2}{3}$ ms and a "make" of $33\frac{1}{3}$ ms. The mechanical design of the dial ensures that there is a pause of at least 200 ms between the pulses generated by successive digits in order to allow switching to be completed by each digit satisfactorily before the next digit signals arrive. The d.c. conditions for signalling digit 2 followed by digit 4 are illustrated in Fig. 5.16.

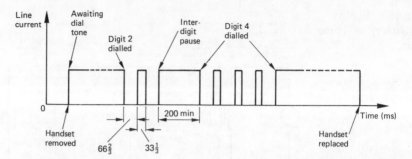

Fig. 5.16 Signalling conditions from rotary dial

It will be appreciated therefore that the automatic exchange switching equipment has been designed to operate to these loop-disconnect pulses which are generated by the dial at a rate of 10 per second.

The dial mechanism also includes electrical contacts which are closed when the dial is moved off its normal position. One contact short-circuits the transmitter to ensure a good d.c. loop condition for signalling and the other contact short-circuits the receiver to prevent loud clicks being heard during dialling.

Magneto bell This responds to the incoming ringing signal from the automatic exchange when the particular instrument telephone number has been dialled. The ringing signal generated at the automatic exchange, as indeed even for the very early manual exchanges, is a low-frequency alternating signal, originally at $16\frac{2}{3}$ Hz with an r.m.s. value of 75 volts. Alternating current is used because a magneto bell is far more reliable and needs far less maintenance than a simple d.c. bell that normally incorporates a break-and-make contact arrangement that very quickly becomes damaged due to "burning" by sparking. The magneto bell is connected in series with a capacitor across the incoming telephone line on the line side of the gravity switch contacts in the instrument.

Induction coil This is essentially a 3-winding auto transformer that provides separate transmission circuits, with correct impedance matching to the line, for the transmitter and receiver respectively. It also forms part of a balance circuit that reduces sidetone, which is the undesirable effect of a talker hearing his or her own voice in the receiver. This problem is due to the choice of a two-wire connection between the telephone instrument and the exchange. A simplified circuit diagram of a standard telephone instrument is shown in Fig. 5.17.

The Electronic Telephone

With all the rapid technological advances in electronic components and circuitry it might be expected that the telephone instrument would have altered considerably from the standard design, in the same way that many other consumer products have done in recent years. It would indeed seem a simple task to produce a modern design incorporating integrated circuitry that could perform the relatively simple functions of the standard instrument. In fact, the design of the standard instrument over the years has become so refined that it is difficult to produce a modern design which is

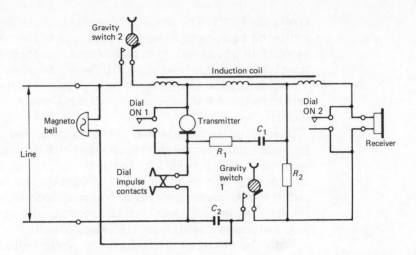

Fig. 5.17 Circuit diagram of typical telephone instrument

significantly cheaper and more reliable for existing exchanges. For example, in Fig. 5.17, capacitor C_2 performs several different functions. It provides a series path with the magneto bell for incoming ringing current; it prevents the flow of direct current in the rocking-armature receiver; and it is also included in the balance impedance and the dial impulsing-spring spark-quench circuits. This is an example of how the instrument design has been optimised to reduce components to a minimum. Also, a smaller and more efficient transmitter inset was introduced a few years ago.

Despite this, however, design of a modern instrument has progressed because an electronic telephone of the future will have to be able to provide a number of advantages in the long term, both to the administration and the user, as the introduction of electronic and digital switching exchanges advances. Production and maintenance costs of the standard instrument have continually increased, largely due to the wire-wound techniques of the bell coils and the induction coil, and the mechanical parts of the dial and the bell assemblies.

Integrated circuits can improve speech transmission, and eliminate noise and distortion associated with the carbon granule transmitter and induction coil. The saving in space provided by the integrated circuits can allow extra facilities to be added which could otherwise only have been achieved by adding extra external units to the standard instrument. Typical extra facilities are storage of certain often-used numbers, display of numbers, call timing and cost indication, and so on.

Two examples of the initial modernisation programme are seen in the Trimphone, where a tone caller replaces the conventional magneto bell, and in the option of replacing the standard dial by a push-button key pad.

Push-button Dialling

As previously explained, the dial is relatively slow in operation, and with long telephone numbers now very common with STD, digit selection errors can be introduced unknowingly by the user. Selection of a number by pressing buttons is simpler and quicker for users, and therefore fewer errors are likely. The problem of course is that existing exchanges have switching equipment designed to respond to the normal dial impulsing speed of 10 per second, so in these exchanges there can be no decrease in time taken to actually set up a call. The push-button keypad operation has to regenerate the same loop-disconnect pulses as the dial, either directly or indirectly, for the exchange switching equipment to function satisfactorily.

In early trials, push-button operation generated unique pairs of audio frequency tones to represent each selected digit, and additional interface equipment was provided at the exchange to convert these tones into normal loop-disconnect impulses to operate the switching equipment. Later, the telephone instrument was modified to store the digits selected by the push-buttons, and then to automatically generate the necessary loop-disconnect impulses for the selected digits.

With the electronic exchanges which are replacing old electromechanical exchanges and with most new designs of PABX, large-scale integrated circuits and microprocessors are used, and the unique audio-frequency tones generated by push-button selection can control the switching equipment directly. This has enabled calls to be set up much faster than in the past, with more efficient use of common equipment.

In early trials the keypad had 10 buttons in two rows of five, but this has since been replaced by 12 buttons in three rows of four. Only ten buttons are needed for digits 0 to 9 inclusive, and the other two have been included and designated "star" and "square" respectively as an agreed international standard for possible future additional user facilities. For multi-frequency signalling, each push-button generates a unique pair of audio frequencies, one from a "low" range and one from a "high" range. The keypad layout and frequency allocations are illustrated in Fig. 5.18.

Fig. 5.18a Layout of push-button keypad

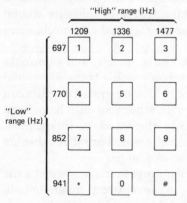

Digit	1	2	3	4	5	6	7	8	9	0	*	#
Audio frequency tones (Hz)	697 1209	697 1336	697 1477	770 1209	770 1336	770 1477	852 1209	852 1336	852 1477	941 1336	941 1209	941 1477

Fig. 5.18b Allocation of tones to individual digits

As a result of extensive development, British Telecom have introduced two new telephone instruments, called the Viscount and the Statesman. These instruments, having a keypad as standard provision, incorporate the latest technology with an integrated circuit for speech transmission and reception in conjunction with smaller low-cost dynamic inset transmitter and receiver. The user will benefit by freedom from high noise levels and distortion inherent in existing instruments.

6 Introduction to Lines, Losses, and Noise

In Chapter 1 the idea of using wires as a line link to convey electronic information signals between two points was introduced as an alternative to using a radio link. It was also seen in Chapters 3 and 4 that, in radio systems, lines are used to carry information signals from radio or tv studio or radio-telephony terminal to a radio transmitting station. Then, in Chapter 5, the idea of using a 2-wire line to connect two telephone instruments together was introduced. Interconnecting lines such as these will now be considered very briefly.

Generally, a *transmission line* can be considered as a conductor, or group of conductors, with suitable insulating materials, whose function is to carry electronic information signals. The line can take various physical forms according to the type of information to be transmitted and the distance involved.

Earth Return Circuits

Early morse code telegraph circuits used a *single* conductor or wire to connect two places together. This is illustrated very simply in Fig. 6.1.

Fig. 6.1 Single wire and earth return telegraph circuit

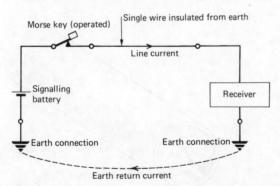

The earth contains large amounts of water and other conductive materials and can be used as a return conductor provided that a good connection with low resistance can be made with it. The main disadvantages of this arrangement, apart from the problem of making a good electrical connection to the earth, are *a* The resistance (or opposition to current flow) of the insulated single wire is greater than the return path through the earth, so the line is unbalanced.

b If other circuits also use the same arrangement, the earth is carrying return currents of all the different circuits, and mutual interference between the various circuits can occur.

c Power supply circuits which themselves do not carry information signals can also produce interference to earth-return circuits.

Two-wire Lines

The disadvantages of the earth-return system can be largely overcome by using two identical conductors insulated from each other and from earth. The two conductors will now have the same resistance, and are not used by any other circuit.

The simplest form of two-wire line is produced by using bare conductors suspended on insulators at the top of poles. This is illustrated in Fig. 6.2.

Fig. 6.2 Simple overhead two-wire line

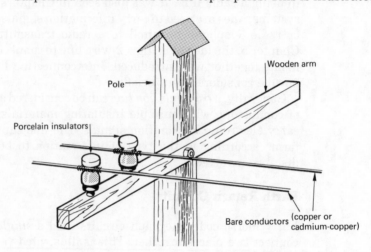

Fig. 6.3 Simple two-wire cable

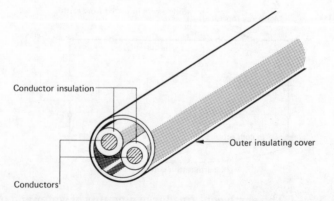

Another type of two-wire line consists of conductors insulated from each other in a cable which also has an outer cover of insulation, as illustrated in Fig. 6.3. Often the two insulated conductors in the cable shown are twisted together along the length of the cable, and are called a *pair*.

Fig. 6.4 Simple illustration of multi-pair cables

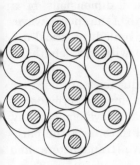

(a)   TWIN-TYPE

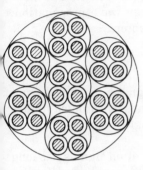

(b)   QUAD-TYPE

Multi-pair Cables

It is often necessary to provide a number of two-wire lines between the same two places, and this is done most conveniently by making a cable with a number of pairs of insulated wires inside it. Sometimes the wires are twisted together in pairs as illustrated in Fig. 6.4a, but sometimes they are provided in fours, or quads, as shown in Fig. 6.4b.

In order to identify the various wires, each one has a colouring on the insulating material around it in accordance with a standard colour code.

Coaxial Cables

As the frequency of an alternating current is increased, the current tends to flow along the outer part of a conductor having a circular cross-section. This means that the centre part of the conductor is not carrying current and can be removed. The empty space can then be used for a second conductor, provided it is insulated from the outer conductor. This type of cable is called a *coaxial cable*, and is illustrated in Fig. 6.5.

The two conductors can be insulated from each other either by a solid insulation along the whole length of the cable, or by insulating *spacers* fitted at regular intervals as supports for the inner conductor. The main insulation in this case is therefore the air between the two conductors.

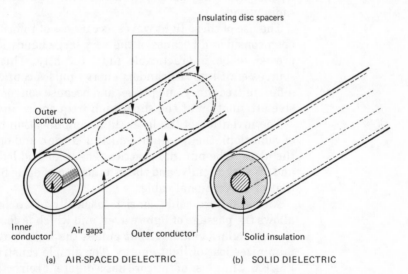

Fig. 6.5 Simple illustration of coaxial cables

Insulating disc spacers

Outer conductor

Inner conductor   Air gaps   Outer conductor   Solid insulation

(a)   AIR-SPACED DIELECTRIC   (b)   SOLID DIELECTRIC

Waveguides

The losses in coaxial cables due to conductor resistance increases as frequency increases. There are also increasing losses due to the insulating materials used to support the inner conductor of the cable. The total losses become very significant in the s.h.f. range 3 to 30 GHz, and are overcome by removing the inner conductor and using a hollow tube to carry the electromagnetic wave, as for gas or water. The electromagnetic wave is launched down the tube by a device similar to an aerial that produces free-space propagation of radio waves. However, the energy is contained within and guided along the tube instead of spreading out as in free space propagation. The tube is therefore called a **waveguide**.

For satisfactory propagation, the cross-sectional dimensions of a waveguide must be directly related to the wavelength of the energy wave. So a waveguide only becomes small enough for convenient practical use when the working frequency reaches the s.h.f. range, with wavelengths in the range 10 to 1 cm.

As previously mentioned in Chapter 3, such frequencies are used for wideband multichannel microwave radio relay systems that provide an important alternative to multichannel coaxial cable systems in any national telephone network. Waveguides are used to connect the parabolic microwave dish aerials to their respective transmitters and receivers.

Waveguides can be rectangular or circular in cross-section, and there are a number of different possible modes of energy wave propagation, discussion of which is beyond the scope of this chapter.

Optic Fibre Cables

Fig. 1.11 indicated that the extensive electromagnetic wave spectrum includes the narrow visible light and infra-red bands, with much higher frequencies, and therefore much smaller wavelengths, than for normal radio waves.

The use of these light waves as carriers of information signals has long been considered because of the very large bandwidth capability, but has proved to be impracticable until recently. This was largely because sources emit light as random energy pulses containing several waves of different frequency and phase, and because conventional cable had excessive attenuation at the super-high frequencies involved.

The invention of the laser (light amplification by stimulated emission of radiation), the led (light-emitting diode), and optic fibre cable enabled the ideas to be put into practice. The laser and led can both be pulsed on and off very rapidly, and the attenuation of optic fibre cable is lower than that of conventional cables.

An optic fibre cable consists basically of an inner glass core, which allows the passage of light waves, and which is completely surrounded by an outer glass cladding that guides the light waves along the core and minimizes loss of light energy. The simple construction is illustrated in Fig. 6.6. The glass of the core has a different refractive index to the glass of the cladding, which results in the light waves being propagated along the inner core by a series of reflections from the outer cladding. The core and the cladding must be free from any impurities to avoid scattering of light energy. Some new optic fibre cables use special plastics instead of glass, resulting in very pure materials with very low attenuation of light signals.

Although lasers and leds are capable of producing visible light outputs, optic fibre telecommunication systems generally use signals in the infra-red band with wavelengths from approximately 0.8 μm to 1.6 μm. Because of the large bandwidths available at these extremely high frequencies,

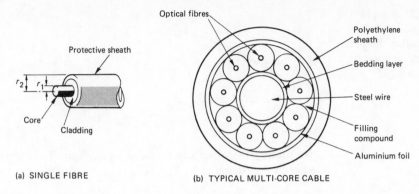

Fig. 6.6 Principles of optic fibre cable construction

Protective sheath

r_2 r_1

Core

Cladding

(a) SINGLE FIBRE

Optical fibres

Polyethylene sheath

Bedding layer

Steel wire

Filling compound

Aluminium foil

(b) TYPICAL MULTI-CORE CABLE

optic fibre cables are particularly suitable for high-capacity digital systems, with bit rates of up to 140 Mbit/s at the present time.

Improving technology is also continually increasing the distances over which such systems can be transmitted before regeneration is needed.

The basic principles of optic fibre telecommunication systems will be considered in more detail in Chapter 15.

Attenuation of Information Signals by Lines

Whatever the type of cable used, the conductors must have some electrical resistance (or opposition to current flow). Furthermore, the insulating material used to separate the two conductors of a pair will have a value of *insulation resistance* which will allow a very small current to flow between the conductors instead of flowing along the conductor to the distant end.

Also, the insulation between the conductors forms a *capacitance* which provides a conducting path between the conductors for alternating currents, the conducting path becoming better as the frequency of the alternating current increases. The capacitance also has the ability to store electrical energy. This capacitive path therefore prevents part of the a.c. information signal from travelling along the conductors to the distant end of the line.

Energy is used up to make the current flow against the resistance along the conductors, and against the insulation resistance between the conductors. Energy is also used in charging and discharging the capacitance between the conductors. In multi-pair cables there is capacitive and inductive coupling between pairs, so that some energy is passed from one pair to other pairs. This reduces the amount of energy that is transmitted along the original pair, and so contributes to the loss.

In the case of an information signal, this energy is extracted from the signal source and so the energy available is gradually decreased as the signal travels along the line. This loss of energy along the line is called **attenuation**. The *unit* used to measure this will be considered later.

If the line is long, and the attenuation is large, eventually the signal energy available at the distant end is too small to operate a receiving transducer. The attenuation generally increases as the frequency of the information signal increases, and this variation of attenuation with

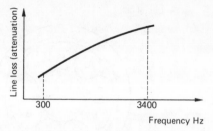

Fig. 6.7 Simple illustration of attenuation distortion of a line at speech frequencies

frequency is called *attenuation distortion*. This is illustrated simply for speech frequencies in Fig. 6.7.

Noise

In any telecommunication system, whether using line or radio links, there is unwanted electrical energy present as well as that of the wanted information signal.

This unwanted electrical energy is generally called **noise** (illustrated in Chapter 1, Fig. 1.1 and Fig. 1.2) and arises from a number of different sources, which will now be considered very briefly.

(1) RESISTOR NOISE

A *conductor* is designed to carry current with minimum opposition, consistent with size and cost.

A *resistor* is a component designed to have a particular opposition to the flow of electric current in a particular circuit. This opposition is called *resistance* in d.c. circuits, but in a.c. circuits the term *impedance* is used because of added factors to be considered later. In either case the *unit* used is the **ohm**(Ω).

An electric current is produced by the movement of *electrons* dislodged by an externally applied voltage from the outer shells of the atoms making up the conductor material or resistor material. The movement or agitation of atoms in conductors and resistors is somewhat random, and is determined by the temperature of the conductor or resistor. The random movement of electrons brought about by thermal agitation of atoms tends to have increased energy as temperature increases.

This random movement of atoms gives rise to an unwanted electrical voltage which is called resistor noise, circuit noise, Johnson noise or thermal noise. This unwanted signal spreads over a wide range of frequencies, and the noise present in a given *bandwidth* required for a particular information signal is very important. It will also be shown later that the important noise temperature of the resistor or conductor is the Absolute or Kelvin temperature, which has its zero point at $-273°$ Centigrade. This is the temperature at which the random movement or agitation of atoms in conducting or resistive materials ceases, so unwanted noise voltages are therefore zero.

(2) SHOT NOISE

This is the name given to noise generated in active devices (energy sources), such as valves and transistors, by the random varying velocity of electron movement under the influence of externally applied potentials or voltages at appropriate terminals or electrodes.

(3) PARTITION NOISE

This occurs in multi-electrode active devices and is due to the total current being divided between the various electrodes.

(4) FLUCTUATION NOISE

This can be *natural* (electric thunderstorms, etc.) or *man-made* (car ignition systems, electrical apparatus, etc.) and again spreads over a wide range of frequencies. Such noise can be picked up by active devices and conductors forming transmission lines.

(5) STATIC

This is the name given to noise encountered in the free-space transmission paths of radio links, and is due mainly to ionospheric storms causing fluctuations of the earth's magnetic field. This form of noise is affected by the rotation of the sun (27.3 day cycle) and by the sunspot activity that prevails.

(6) COSMIC OR GALACTIC NOISE

This type of noise is also most troublesome to radio links, and is mainly due to nuclear disturbances in all the galaxies of the universe.

(7) In multi-pair cables there is capacitive and inductive coupling between different pairs which produces an unwanted noise signal on any pair because signals are transmitted to other pairs. This is called CROSSTALK between pairs and can be reduced to some extent by twisting the conductors of each pair or by changing the relative positions of pairs along the cable during manufacture or by balancing the pairs over a particular route after installation.

(8) FLICKER NOISE

The cause of this is not well understood but it is noise which predominates at low frequencies below 1 kHz, with the level decreasing as frequency increases. It is sometimes known as "excess noise" or "$1/f$ noise."

In any telecommunications system, therefore, there will be a certain level of noise power arising from all or some of the sources described, with the noise power generally being of a reasonably steady mean level, except for some noise arising from *impulsive* sources such as car ignition systems and lightning. Noise which has a sensibly constant mean level over a particular frequency bandwidth is generally called *white noise*.

In order that a wanted information signal can be detected and reproduced satisfactorily at the receiving end of a system, it is essential that the power of the wanted signal is *greater* than the noise power present by at least a specified minimum value. This introduces the very important concept of **signal-to-noise ratio** in any telecommunication system as the comparison of signal power to noise power. It can be expressed simply as a *power ratio*, or more commonly it is expressed in *decibels* (dB).

The derivation of this important unit as a logarithmic ratio will not be dealt with here but it can be simply stated as

$$\text{Signal-to-Noise ratio} = 10 \log_{10}\left(\frac{\text{Signal power}}{\text{Noise power}}\right) \text{ decibels}$$

For any type of information signal there is a *minimum* acceptable value of signal-to-noise ratio for the system to operate satisfactorily. Typical *minimum* signal-to-noise ratios for different systems are as follows:

1 Private land mobile radio telephone systems require 10 dB.
2 Ship-to-shore radio telephone services require 20 dB.
3 Telephone calls over the public network require 35 to 40 dB.
4 Television systems require 50 dB.

Now, returning to the problem of sending an information signal along a line, transistor amplifiers can be used to increase the signal level to compensate for the attenuation of the line. Each amplifier will generate noise internally, as previously described, so the output of each amplifier will contain the wanted signal and unwanted noise with a certain signal-to-noise ratio.

There will be also be Johnson noise present on the line because of the resistance of the line conductors, and also crosstalk noise from other lines.

One amplifier *could* be placed at the sending end as shown in Fig. 6.8*a* with sufficient amplifying properties or *gain* to compensate for the line attenuation, so that the information signal reaching the other end of the line has sufficient power to operate the receiving transducer satisfactorily. This could result in a large signal power at the sending end which would cause excessive interference to other circuits in the same cable due to mutual inductance and capacitive coupling between different pairs. To avoid this problem there is a maximum permissible signal power laid down for application to pairs in different types of cable.

Another way to overcome attenuation would be to put *one* amplifier at the receiving end as shown in Fig. 6.8*b* with sufficient gain to compensate for the line attenuation. However, if the line is long with a restricted permissible power at the sending end, the attenuation could be such that the information signal power at the receiver is low enough to give an inadequate signal-to-noise ratio when the line noise and noise generated by the receiver are considered.

To overcome these problems, amplifiers must be placed at regular points along the line where the information signal power is still large enough to give an adequate signal-to-noise ratio compared with the amplifier noise and line noise. This simple concept is illustrated in Fig. 6.8*c*.

Since an amplifier is generally a one-way device with definite input and output connections, the arrangement illustrated in Fig. 6.8 needs to be duplicated to enable information signals to be transmitted in the opposite direction. However it has previously been seen that simple telephone

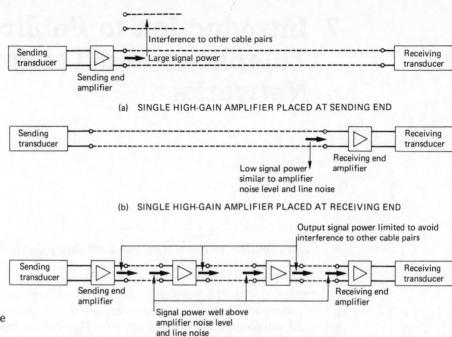

(a) SINGLE HIGH-GAIN AMPLIFIER PLACED AT SENDING END

(b) SINGLE HIGH-GAIN AMPLIFIER PLACED AT RECEIVING END

(c) LIMITED-GAIN AMPLIFIERS PLACED AT REGULAR INTERVALS ALONG A LINE

Fig. 6.8 Various ways of using amplifiers to overcome line attenuation in a one-way tele-communication system

communication circuits carry information in both directions over a single pair of wires. To meet this requirement, it is therefore necessary to arrange that when amplification is needed over telephone circuits, the normal simple two-wire connection is changed into a 4-wire connection to provide one pair for transmitting signals in each direction. This arrangement will be dealt with in more detail later, but the simple principles are illustrated in Fig. 6.9.

It should be added here that there are certain types of amplifier that can be inserted into a 2-wire line to give amplification in both directions, but the use of these in the public telephone network is limited.

Fig. 6.9 Illustration of 4-wire amplified telephone circuit

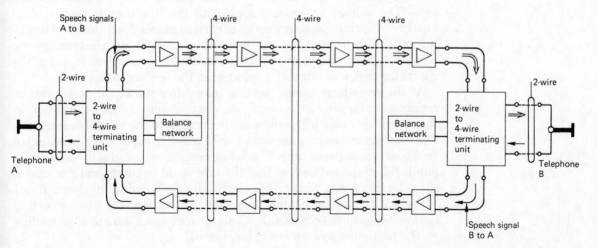

7 Introduction to Public Telephone and Telegraph Networks

Introduction to Networks and Switching Centres

Fig. 5.2 (p. 43) illustrated very simply the principle of a telephone connection, with the telephone instrument at each end of the link having sending and receiving transducers known as the telephone transmitter and telephone receiver respectively. This can be represented even more simply as shown in Fig. 7.1.

Fig. 7.1 Simple telephone system

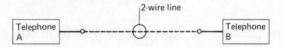

Another form of communication system, which was mentioned in Chapter 1, deals with the written word instead of the spoken word, and is called a telegraph system. One type of telegraph system has two teleprinters connected together by a line.

A **teleprinter** is a device that looks like a large typewriter, having a keyboard with letters, figures and other commonly-used characters. When any key is depressed, the teleprinter mechanism produces an electrical signal that represents the particular character in the form of a series of voltage pulses at +80 V and −80 V in accordance with a 5-unit or Murray code. The voltage pulses are applied to the line that is connected to another teleprinter so that a series of current pulses flows along the line to the other teleprinter. The current pulses energise an electromagnetic receiver which then operates the teleprinter mechanism to print the character that was originally selected at the sending teleprinter.

At the same time, at the sending teleprinter the signal pulses can be connected internally to the receiving electromagnet of the teleprinter, so that a copy is made of the message being sent to the distant teleprinter. The message can be produced on a normal sheet of paper (page-printing) or on a continuous paper tape. The latter method was originally used in the public telegraph network so that the tape could be cut up and stuck on to the telegram forms which were eventually delivered to the destination.

Fig. 7.2 illustrates a simple teleprinter connection. The line connecting the two teleprinters can be either a single wire or a 2-wire line, depending on the particular system being considered.

Fig. 7.2 Simple telegraph system

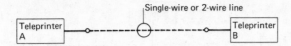

The two communication systems illustrated in Figs. 7.1 and 7.2 show permanent connections between two points A and B, and represent the very early uses of these two methods of communication. It is not difficult to imagine that, in the very early days of telephone and telegraph communication, other people would want to be included in such systems, and that each person or *subscriber* to a system would wish to have access to all other subscribers when necessary.

If we consider a small number, say four or five, wishing to set up a telephone or telegraph communication system, it is possible to arrange complete interconnection by providing suitable line connections between all the telephone or telegraph subscribers as shown in Fig. 7.3. In such a small system it is necessary to provide some form of signalling code so that each subscriber can call any other subscriber when desired.

Fig. 7.3 Fully interconnected communication system

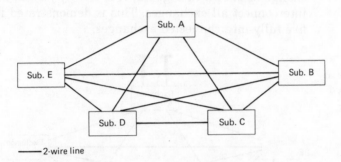

It should also be fairly obvious that there is a limit to the size of such a fully-interconnected system, both in terms of number of subscribers and in the geographical area that can be covered. One can imagine the problems of connecting one subscriber to hundreds of thousands of other subscribers over long distances.

It was a logical development therefore to provide a central point to which all subscribers are connected, and where any two subscribers are interconnected on demand. This central point is called an **exchange**, because connections can be exchanged when required, and we therefore see the introduction of telephone and telegraph exchanges as switching centres, with operators employed to do the switching on demand by calling subscribers.

The principle of a telephone exchange system is illustrated in Fig. 7.4.

Once a telephone or telegraph system had been set up in a particular country, many other subscribers would wish to be included, and sooner or later the problem of providing a national network would arise. Considering Fig. 7.4 again, it should be clear that it would be impracticable to provide *one* central telephone or telegraph exchange for a country with all subscribers connected to it, because of the amount of switching equipment that would be necessary, and because of the length of line needed for each

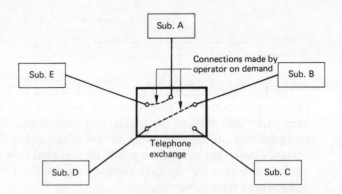

Fig. 7.4 Simple telephone system with a central switching point or exchange

Sub. A

Sub. E

Connections made by operator on demand

Sub. B

Telephone exchange

Sub. D

Sub. C

subscriber. So, a large number of exchanges are needed in a **national network**, with each exchange having clearly defined geographical boundaries and a limited number of subscribers connected to it.

The problem then arises as to how to enable subscribers on one telephone exchange to be connected to any other subscriber in the country. Just as it is impracticable to fully-interconnect subscribers on one exchange as shown in Fig. 7.3, it is clearly just as impracticable to fully-interconnect all exchanges. This is demonstrated in Fig. 7.5, with only five fully-interconnected exchanges.

Fig. 7.5 Five telephone exchanges in a fully interconnected group

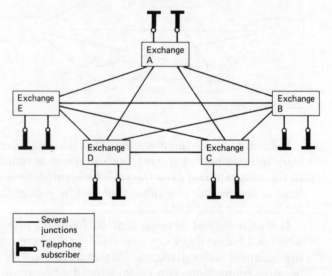

Exchange A

Exchange E

Exchange B

Exchange D

Exchange C

Several junctions

Telephone subscriber

The lines interconnecting the exchanges represent a sufficient number of 2-wire links to handle the calls between subscribers from any two exchanges. These interconnecting lines are called **junctions**. Clearly if this idea was extended to a large number of exchanges, the number of lines needed would be immense. This problem can be tackled by arranging telephone exchanges in a particular geographical area into a *group* of exchanges, and selecting *one* of the exchanges as a switching centre for the group and called a **group switching centre** (GSC). This is illustrated in Fig. 7.6.

Connections can be made within a group of local exchanges by a fully-interconnected junction network, but all connections to subscribers *out-*

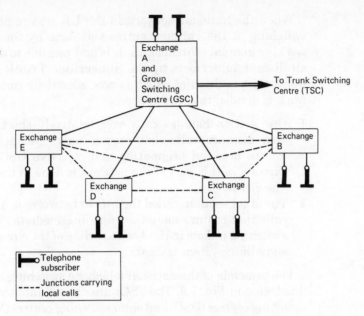

Fig. 7.6 Concentration of trunk calls at a group switching centre

Exchange A and Group Switching Centre (GSC)

To Trunk Switching Centre (TSC)

Exchange E

Exchange B

Exchange D

Exchange C

○┱ Telephone subscriber

- - - - Junctions carrying local calls

side the group are routed via the GSC. In this way a number of local exchanges share one GSC, and calls outside the group are therefore concentrated at the GSC.

In the same way, a number of groups can share one switching centre, for routing long-distance calls, as illustrated in Fig. 7.7. The switching centres for long-distance trunk calls are called **trunk switching centres** (TSC). Also shown in Fig. 7.7 are *direct* routes between adjacent GSCs where the number of calls justifies such provision.

In this way, by setting up a graded network or hierarchy of switching centres, the whole of a country can be included with a minimum number of interconnecting junctions and trunk circuits, with calls being collected or concentrated at strategic points.

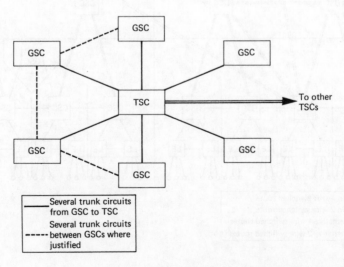

Fig. 7.7 Concentration of trunk calls at a trunk switching centre

GSC

GSC

GSC

TSC

To other TSCs

GSC

GSC

GSC

GSC

——— Several trunk circuits from GSC to TSC

- - - - Several trunk circuits between GSCs where justified

When the national network in the UK was re-planned in the 1930s, switching at the various centres was done by the telephone operators working manual switchboards. It is now possible to dial directly to nearly all distant subscribers by the **Subscriber Trunk Dialling** (STD) network, the provision of which is now effectively complete. The STD network is divided into two sections:

1   One section handles calls over relatively short distances, and also long-distance calls that require only one intermediate switching centre between originating and objective GSCs. The switching of circuits in this part of the network is done in the two-wire portion of the circuits.

2   The other section, called the *transit* network, is used for long-distance calls that require more than one intermediate switching centre. The switching is done in the 4-wire portion of the circuits, and a high-speed signalling system is used.

The principle of the national telephone switching network in the UK is illustrated in Fig. 7.8. The TSCs are divided into two categories, *district switching centres* (DSC) and *main switching centres* (MSC), with the MSCs being fully interconnected. Each GSC is connected to at least *one* DSC.

So, in the national network there are various grades of exchange in order of importance. The more important a particular grade of exchange is, the fewer there are. There are also various grades of interconnecting line, for example subscribers' lines, junctions and trunks, and again the number provided gets fewer as length, importance and cost increase.

Fig. 7.8   Outline of a national telephone switching network

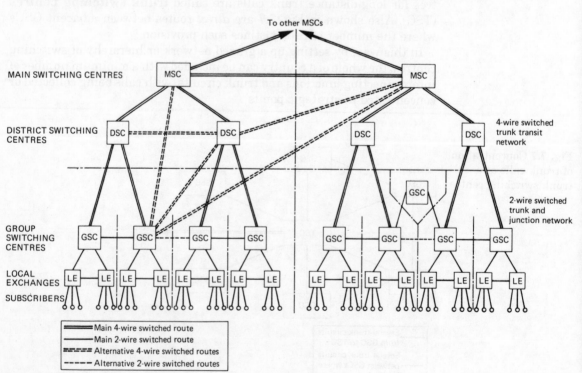

Fig. 7.9 Typical connection between two subscribers via transit network

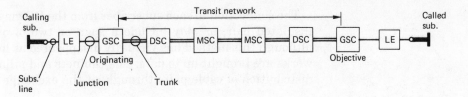

Fig. 7.10 Connection between two subscribers without using transit network

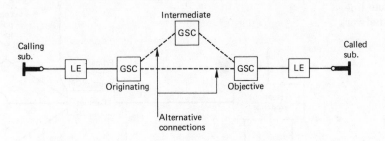

Fig. 7.11 Typical possible connections from subscribers to international exchange

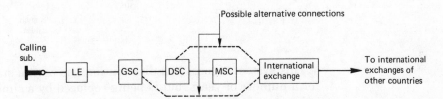

An illustration of how two telephone subscribers in different parts of the country could be connected together is given in Fig. 7.9.

For calls between two subscribers relatively close to each other, the transit network is not needed, as shown in Fig. 7.10. The intermediate GSC may or may not be needed, according to the distance between subscribers.

The next requirement for a subscriber may be to make a telephone call to another country. In the same way as national calls are progressively concentrated at GSCs, DSCs and MSCs, so calls to other countries are concentrated at an **international exchange** in any country. In the UK the international exchange is located in London. Fig. 7.11 illustrates a connection to the international exchange.

Local Distribution Network

The way in which subscribers' telephone instruments are connected to the local telephone exchange will now be considered, bearing in mind that each telephone requires a two-wire line or **pair**.

These pairs leave the exchange in large multi-pair *main* cables which feed primary connection points (PCP) or cast-iron *cabinets* placed at various points in the telephone exchange area.

Each PCP or cabinet is then connected to a number of secondary connection points (SCP) or *pillars* by means of smaller multi-pair *branch* cables. From the SCPs or pillars, distribution cables are connected to *distribution points* (DP), each of which will feed a small number of subscribers' premises.

The connection to each subscriber from the DP may be a simple drop-off insulated pair, or *may* include a span or two of overhead bare wires, although this method has virtually disappeared as local distribution networks are brought up to date. The cabinets and pillars give flexibility in distribution of cable pairs throughout the exchange area.

Fig. 7.12 Principle of local telephone line distribution network

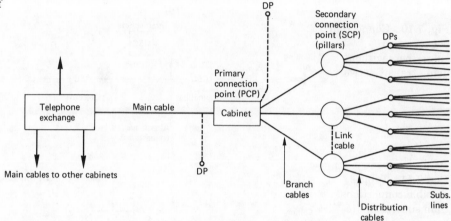

This system of local line distribution has been a standard installation for a number of years, but is being replaced by a similar flexible system using cast-iron cabinets instead of the original concrete-type pillars for the SCPs, with a different method of connecting pairs together inside the cabinets. This is illustrated in Fig. 7.12.

Connections between Telephone Exchanges

In Fig. 7.8 the national network is illustrated, with local subscribers' lines, junctions between certain exchanges, and trunk circuits between other exchanges. In Chapter 6 the different types of cable were mentioned briefly.

Generally, for the low-category lines from subscribers to local exchanges, and for junctions between local exchanges and GSCs, **multi-pair** types of cable are used. Between GSCs and TSCs, and between TSCs, where many originating calls have been concentrated, **coaxial**-type cables will generally be used.

These cables are generally buried in the ground, but occasionally submarine cables may be required where the cable has to cross rivers, lakes or estuaries, for example.

Supervisory Signals

When national telephone networks were first introduced, the switching between lines was done by telephone operators working at manual switchboards. It was soon realized that there are several advantages in providing automatic equipment which is remotely controlled by calling subscribers using a dial incorporated in the telephone instrument.

In the manual system, when a subscriber wishes to make a call, a signal is sent to call the operator. The calling subscriber knows when the operator has answered because the operator says "number please". In the automatic system, some distinctive form of signal must be passed to the calling subscriber to state that the automatic equipment is ready to receive routing information from the dial. This is called **dial tone**.

In the manual system, when the operator *says* "number please", the calling sub gives the number required. The operator tests the wanted line, and if it is free, sends a ringing current to ring the bell of the wanted telephone, and *tells* the calling sub: "trying to connect your call". To convey the same information in an automatic system a recognizable signal is needed, and this is called **ring tone**. At the same time *ringing current* is applied to the called subscribers line to ring the telephone bell, just as the operator does in the manual system.

In the manual system, if the wanted subscriber is already engaged on a call, the operator *tells* the calling subscriber that the line is engaged. In an automatic system, a **busy tone** is needed to give the same information. If a calling subscriber asks the operator in a manual system for a number that is out of order, or is not a working line, then the calling subscriber is *told* that this is so. In an automatic system a **number unobtainable** (NU) **tone** is used to convey this information to the calling subscriber.

In an automatic exchange, when a calling subscriber dials a wanted number, the automatic equipment *may* not be able to connect the call because certain sections may all be engaged on other calls. This information must be conveyed to the calling subscriber by means of an **equipment busy tone**.

Telegraph Networks

In most countries a need emerged for two teleprinter networks. One is a *public* network that enables people to send what used to be called *telegrams* to any address in the country. Such telegrams were originated either by handing in a *written* telegram form at a Post Office, or by *telephoning* the telegram to a special telegraph switchboard operator by dialling a particular code. The telegram message was then conveyed by teleprinter to the telegraph terminal nearest the destination. The telegraph operator then stuck the teleprinter tape message on to a telegram form, which was then delivered to the destination address.

The second need for a teleprinter network arises from private communication between individuals or firms and organizations where a *printed record* of the communication is preferred to a *spoken* message by telephone. Such a system is known as **telex**.

In the early days of private teleprinter communication, the public telephone network was used. Each subscriber who needed the facility rented a teleprinter as well as a telephone. To send a Telex message, a telephone call was made to the wanted number, and when the connection was established both subscribers switched from telephone to teleprinter, and the message was transmitted. This system is illustrated in Fig. 7.13.

The voice-frequency equipment shown in Fig. 7.13 is needed to change the ±80 V coded d.c. pulses into alternating voltage having a frequency within the commercial speech range, so that the telephone line can handle the teleprinter information signals. Several problems arose from this

Fig. 7.13 Original arrangement of telex network

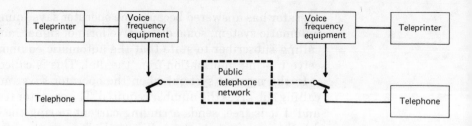

Fig. 7.14 Automatically switched telex network

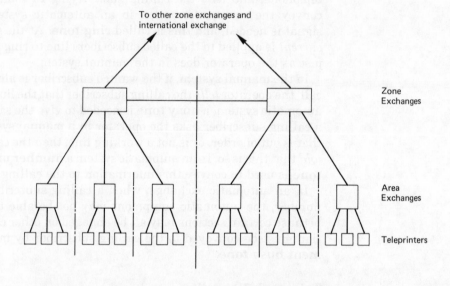

arrangement, and it was decided to set up a completely separate switching network for the Telex Service, which was completed in the UK in 1954.

This separate Telex network originally used exchanges with manual switching by Telex operators, but these exchanges were eventually replaced by approximately 50 automatic Telex exchanges. Each Telex exchange serves a certain area, and a number of areas form a Zone. There are six zones, which are fully interconnected. The network is illustrated in Fig. 7.14.

8 Introduction to Telephone Exchange Switching Principles

In Chapter 7 the need for telephone and telegraph exchanges in a national switching network was introduced. We now have to consider some different ways in which the switching between lines is achieved in exchanges.

Matrix Switching

One method is to use a **matrix switch**, whose principle can be explained by considering the circuits which are to be connected together as being arranged at right angles to each other in horizontal and vertical lines. These lines represent inlets and outlets of the switch. This idea is illustrated in Fig. 8.1.

Fig. 8.1 Simple 4 × 4 matrix switch (see Fig. 8.8 for typical crosspoint connections)

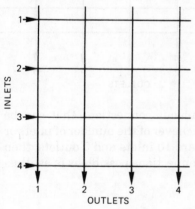
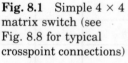

Fig. 8.2 Simple principles of switching by a 4 × 4 matrix switch

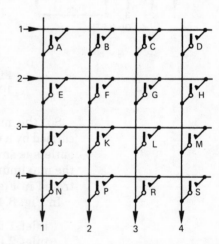

The intersections between horizontal and vertical lines are called **crosspoints**. At each crosspoint some form of switch contact is needed to complete the connection between horizontal and vertical line, as shown in Fig. 8.2.

Any of the 4 inlets can be connected to any of the 4 outlets by closing the appropriate switch contacts. For example,

a Inlet 1 can be connected to outlet 2 by closing contact B.
b Inlet 4 can be connected to outlet 3 by closing contact R.

Considering Figs. 8.1 and 8.2 again, it can be seen that with 4 inlets and 4 outlets there are 16 crosspoints. Obviously, the number of crosspoints in any matrix switch can be calculated by multiplying the number of inlets by the number of outlets. This is further illustrated in Fig. 8.3.

If there are n inlets and m outlets, then the number of crosspoints is $(n \times m)$.

1 If n is larger than m, that is if there are more inlets than outlets, then not all the inlets can be connected to a different outlet. When all the outlets have been taken there will be some inlets still not in use.

2 If m is larger than n, that is there are more outlets than inlets, then when all inlets are each connected to an outlet, there will be some outlets still not in use.

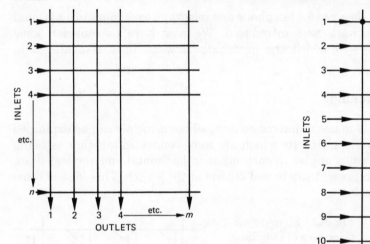

Fig. 8.3 Number of crosspoints in a matrix switch

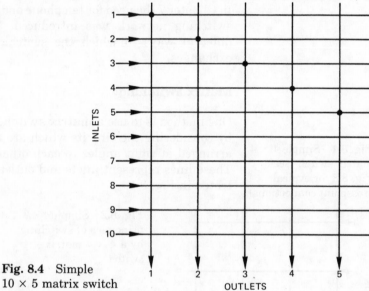

Fig. 8.4 Simple 10 × 5 matrix switch

So, the maximum number of simultaneous connections that can be carried by a matrix switch is given by whichever of the number of inlets or outlets is smaller. For example, if there are 10 inlets and 5 outlets, then the maximum number of simultaneous connections possible is 5, as illustrated in Fig. 8.4.

In Fig. 8.4,

    inlet 1 is connected to outlet 1
    inlet 2 is connected to outlet 2
    inlet 3 is connected to outlet 3
    inlet 4 is connected to outlet 4
    inlet 5 is connected to outlet 5

Efficiency The suitability of a matrix switch as previously described is sometimes measured in terms of the efficiency in the use of its crosspoints. Take a simple matrix switch with 4 inlets and 4 outlets as illustrated in Fig. 8.5. There are 16 crosspoints, but only 4 can be in use at any one time, when the 4 inlets are connected to the 4 outlets.

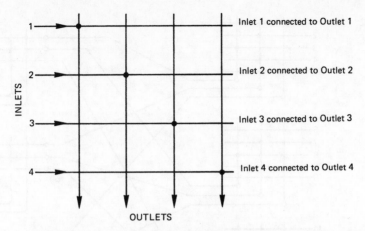

Fig. 8.5 Maximum number of simultaneous connections in a 4 × 4 matrix switch

INLETS

OUTLETS

Inlet 1 connected to Outlet 1

Inlet 2 connected to Outlet 2

Inlet 3 connected to Outlet 3

Inlet 4 connected to Outlet 4

The efficiency in use of crosspoints of the matrix switch is calculated by

$$\frac{\text{Maximum number of crosspoints in use simultaneously}}{\text{Total number of crosspoints in the matrix}} \times 100\%$$

In the case shown in Fig. 8.5,

$$\text{Efficiency} = \frac{4}{16} \times 100\% = \frac{100}{4}\% = 25\%$$

The efficiency of this type of matrix switch gets smaller as the switch gets larger. For example, consider a matrix switch with 15 inlets and 15 outlets, as shown in Fig. 8.6.

The total number of crosspoints = 15 × 15 = 225.

Maximum number of crosspoints in use simultaneously is 15.

$$\text{Efficiency} = \frac{15}{225} \times 100\% = \frac{15}{2.25}\% = 6.67\%$$

Fig. 8.6 15 × 15 matrix switch with 225 crosspoints

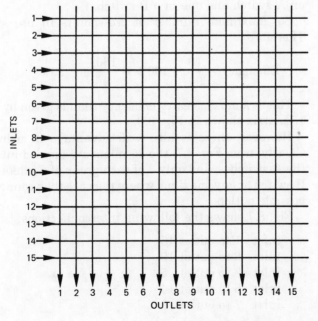

INLETS

OUTLETS

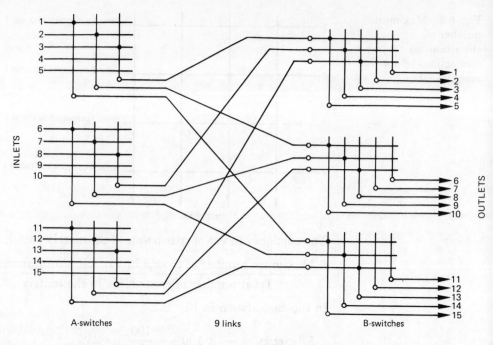

INLETS

OUTLETS

A-switches   9 links   B-switches

Fig. 8.7 Two-stage 15 × 15 matrix switch using link trunking

This low efficiency can be improved for the same number of inlets and outlets by arranging the switch in *two* stages instead of one, using small basic matrix switches. This is illustrated in Fig. 8.7, where six 5 × 3 basic matrix switches are arranged to give a two-stage 15 × 15 switch. The two stages are connected together by 9 links.

In Fig. 8.7, the total number of crosspoints is 6 × 5 × 3 = 90.

Since there are only 9 links between the A and B switches, only 9 inlets can be connected to 9 outlets at any one time.

Each connection from an inlet to an outlet uses *two* crosspoints, one in an A-switch and one in a B-switch.

So, maximum number of crosspoints in use simultaneously is 18. Therefore

$$\text{Efficiency} = \frac{18}{90} \times 100\% = \frac{180}{9}\% = 20\%$$

This is a much greater efficiency than is given by the single-stage 15 × 15 matrix shown in Fig. 8.6.

However, there is also a disadvantage in using this arrangement. Clearly, from Fig. 8.7, the 15 inlets are divided into three groups of 5 at the A-switches, and only 3 of each group of 5 inlets can be connected to a B-switch at any one time, where they are then connected to any 3 of the 15 possible outlets.

Fig. 8.7 shows the following interconnections:

inlet 1 to outlet 15
inlet 2 to outlet 10
inlet 3 to outlet 5
inlet 6 to outlet 14
inlet 7 to outlet 9

inlet 8 to outlet 4
inlet 11 to outlet 13
inlet 12 to outlet 8
inlet 13 to outlet 3

So, inlets 4, 5, 9, 10, 14 and 15 cannot be connected to any outlet, even though outlets 1, 2, 6, 7, 11 and 12 are free, because all 9 links are already in use.

Further, even with only *one* connection made, say between inlet 1 and outlet 15 as shown, then inlets 2, 3, 4 and 5 cannot be connected to any of the outlets 11, 12, 13 or 14 because the *one link* between the particular A and B switches is already in use.

This is called **internal blocking**, or **link congestion** and must be considered when designing multi-stage matrix switches.

In telephone exchanges it is necessary to connect the two wires from one telephone to the two wires of another telephone. It is also necessary to be able to *guard* the calling line and the called line so that neither can be seized by another subscriber. Generally a third wire called the **private wire** and designated P-wire is used for this purpose. There may also be need for a fourth wire or connection for switching and holding purposes. So, each individual *crosspoint* in a matrix may consist of 4 input wires to be connected to 4 output wires.

This is illustrated in Fig. 8.8 which shows a typical crosspoint arrangement in a TXE.2 electronic-controlled exchange that uses reed relays as the crosspoint switches. (A brief description of a reed relay is given later.)

To connect inlet 1 to outlet 1, a current is passed through the reed relay coil (RL), and the 4 contacts operated by the reed relay connect the $-$, $+$, P and H wires of the inlet to the $-$, $+$, P and H wires of the outlet respectively.

Fig. 8.8 Typical application of matrix switch crosspoint connection

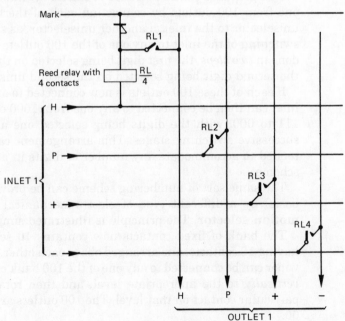

Step-by-Step Switching

This is another method of exchange switching used for many years in several countries, including the UK. The selection of a particular line is based on a one-from-ten selection process. For example, Fig. 8.9 shows a simple switch that has ten contacts arranged around a semi-circular arc or *bank*, with a rotating contact arm or *wiper* that can be made to connect the inlet to any one of the ten bank contact outlets as required. The wiper

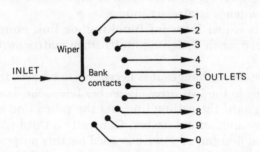

Fig. 8.9 Simple principle of switching by electro-mechanical uniselector of one-from-ten outlets

is rotated by a simple electro-magnet driving a suitable mechanism, so the arrangement is called an *electro-mechanical switch*.

The wiper rotates in one direction only, so this type of electro-mechanical switch is called a **uniselector**. Clearly the inlet can be connected to any one of the ten outlets, but the outlets are numbered from 1 to 0, which is normal practice in the step-by-step switching system.

This principle can be extended to enable the inlet to be connected to any one from 100 outlets by connecting each of the ten outlets of the first uniselector to the inlet of another uniselector, as shown in Fig. 8.10. The switching of the inlet to any one of the 100 outlets (numbered 11 to 00) is done in *two steps*, the first digit being selected on the first uniselector, and the second digit being selected on the second uniselector.

If each of these 100 outlets is now connected to another uniselector, the inlet can then be connected to any one from 1000 outlets, numbered from 111 to 000, with the digits being selected one at a time on the three successive switching stages. This arrangement can theoretically be extended to accommodate any number of digits in a particular numbering scheme.

The same sort of numbering scheme can be provided (on a step-by-step basis) by a different type of electro-mechanical switch called a **two-motion selector**. The principle is illustrated simply in Fig. 8.11.

The bank of fixed contacts now contains 10 semi-circular arcs, each having 10 contacts, and arranged above each other. The moving contact or wiper can be connected to any one of the 100 bank contacts by first moving *vertically* to the appropriate level, and then rotating *horizontally* to a particular contact on that level. The 100 outlets are numbered from 11 to 00.

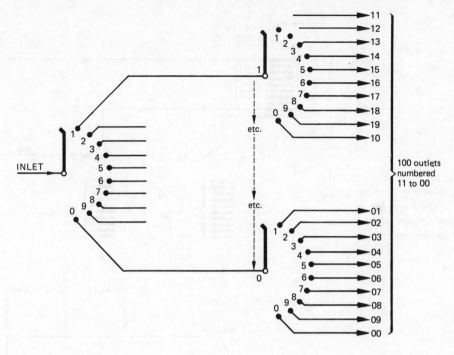

Fig. 8.10 Simple step-by-step selection of one-from-a-hundred outlets

INLET

etc.

etc.

100 outlets numbered 11 to 00

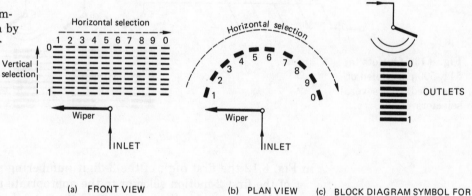

Fig. 8.11 Simple principle of one-from-a-hundred selection by two-motion selector

Horizontal selection

Vertical selection

Wiper

INLET

(a) FRONT VIEW

Horizontal selection

Wiper

INLET

(b) PLAN VIEW

INLET

OUTLETS

(c) BLOCK DIAGRAM SYMBOL FOR TWO-MOTION SELECTOR

The diagram symbol used to illustrate the 100-outlet 2-motion selector is shown in Fig. 8.11c.

As with the uniselector arrangement, the two-motion selector system can be extended to give access to any number of outlets by adding an extra switching stage for each extra digit required in the numbering scheme. A 3-digit numbering scheme from 111 to 000 outlets is illustrated in Fig. 8.12, with the one-from-a-hundred selector preceded by a one-from-ten selector.

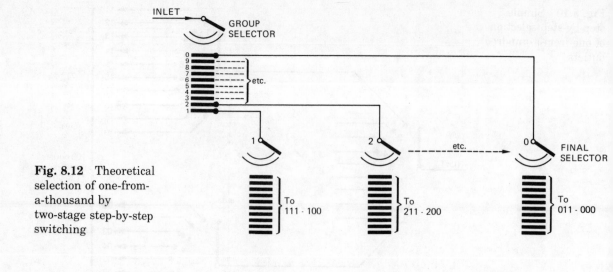

Fig. 8.12 Theoretical selection of one-from-a-thousand by two-stage step-by-step switching

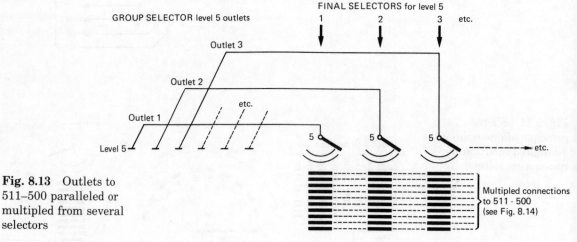

Fig. 8.13 Outlets to 511–500 paralleled or multipled from several selectors

In Fig. 8.12 the first digit of the 3-digit numbering scheme raises the *wiper* of the first 2-motion selector to the appropriate *vertical* level. The selector then automatically searches for a *free* outlet *on that level* to the next selector which caters for the last *two* digits of the 3-digit numbering scheme, as illustrated in Fig. 8.11.

The arrangement of the outlets from *one* level of the first stage of selection is illustrated in Fig. 8.13. The second stage of switches connected from level 5 of the first stage *shares* the 100 possible outlets 511 to 500. The method of sharing or paralleling the outlets from a number of selectors in a *multiple* is illustrated in Fig. 8.14.

The two-motion selectors of the first stage, which select a particular *level* according to the *first* digit of a 3-digit numbering scheme, are called **group selectors**.

The two-motion selectors of the *second* stage which handle the last *two* digits of a 3-digit numbering scheme are called **final selectors**.

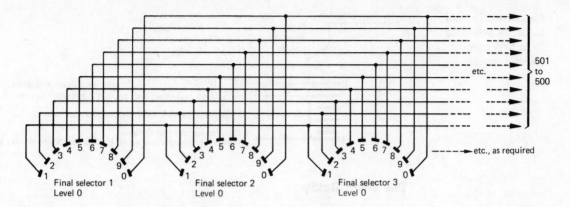

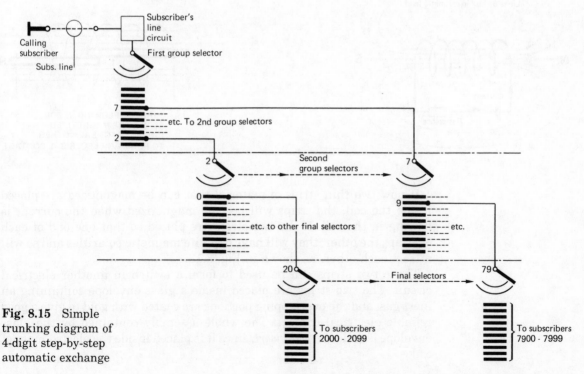

Fig. 8.14 Illustration of multipled outlets from level 0 of Final selectors in Fig. 8.13 (Final selectors connected from level 5 of group selectors as in Fig. 8.13)

In order to provide access to 10 000 lines a further stage of group selectors is added *before* the final selectors. Fig. 8.15 illustrates how a calling subscriber can be connected to other subscribers in an exchange having a 4-digit numbering scheme.

Theoretically, a 4-digit numbering scheme can accommodate 10 000 subscribers, but it is necessary also to provide junctions to other exchanges, junctions to the GSC for STD calls, lines to the operator and other Enquiry Services, lines to telegrams, and so on. This means that only levels 2, 3, 4, 5, 6 and 7 are usually available on the first group selectors for connection to other subscribers via second group selectors and final selectors. So the capacity of the exchange is reduced to 6000 subscribers instead of the theoretical number of 10 000.

The usual arrangement for the first group selector levels is shown in Fig. 8.16.

Fig. 8.15 Simple trunking diagram of 4-digit step-by-step automatic exchange

Fig. 8.16 Typical facilities available from first group selector levels

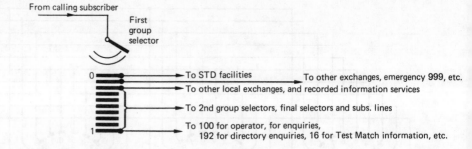

From calling subscriber

First group selector

0
To STD facilities   To other exchanges, emergency 999, etc.
To other local exchanges, and recorded information services
To 2nd group selectors, final selectors and subs. lines
To 100 for operator, for enquiries,
192 for directory enquiries, 16 for Test Match information, etc.
1

The Reed Relay

A relay is a device for remotely operating switches to control the flow of current in other electrical circuits. A reed relay is one particular type of relay used amongst other applications in the control circuits of electronic telephone exchanges. It is based on the fact that an electric current passing through a coil of wire produces an electro-magnet, with the ends of the coil having opposite magnetic polarities, as illustrated in Fig. 8.17.

Fig. 8.17 Coil of wire as a simple electromagnet

Fig. 8.18 Principle of operation of a reed relay

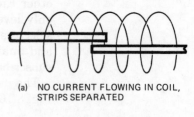

(a)   NO CURRENT FLOWING IN COIL, STRIPS SEPARATED

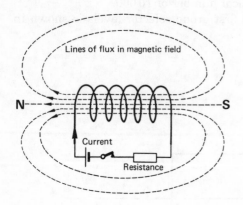

Lines of flux in magnetic field

N — S

Current

Resistance

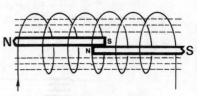

N   S

(b)   CURRENT FLOWING IN COIL, STRIPS ARE MAGNETIZED AND ATTRACT EACH OTHER TO FORM AN ELECTRICAL CONTACT

If now two thin strips of material that can be magnetized are placed inside the coil, the strips will become magnetized when the current is flowing in the coil. If the two strips are placed so that one end of each overlaps the other, they will have opposite magnetic polarities and so will attract each other, as shown in Fig. 8.18.

These two strips can be used to form a switch in another electrical circuit. The two strips are placed inside a glass envelope containing an inert gas, and the overlapping portions are coated with gold to give a good reliable electrical contact. The whole assembly contained by the glass envelope is called a *reed insert*, since it is placed inside the electro-magnet coil.

A typical reed relay, as illustrated in Fig. 8.18, has *four* of these reed inserts placed inside the electro-magnet coil, and each of these can be used to switch a separate electrical circuit.

Individual (or Distributed) Control

Step-by-step automatic switching, or **Strowger switching** as it is called after its American inventor, involves the selection of a dialled telephone number one digit at a time by a series of electro-mechanical two-motion selectors. The number of selectors needed depends on the number of digits used in a particular exchange numbering scheme. Each selector has its own control equipment consisting of electro-magnetic relays and electro-magnets to operate the switch mechanism. The shaft which carries the moving wipers is raised vertically and then twisted horizontally to select a particular outlet. Once the wipers are connected to a selected outlet, the control equipment is idle until the selector is released from the connection and is available to deal with another call.

The total control equipment of an exchange is therefore distributed throughout all the individual selectors, with each selector having exclusive use of its own control equipment – hence the name *individual* (or *distributed*) *control*.

A typical 4-digit call set-up in a Strowger exchange is illustrated in Fig. 8.19, and it should be clear that the route through the exchange switching equipment is selected one digit at a time. The switching of one digit has to be completed before the next digit arrives, and this established the operating speed of the conventional rotating dial fitted to each

Fig. 8.19 Principle of 4-unit step-by-step (Strowger) exchange switching

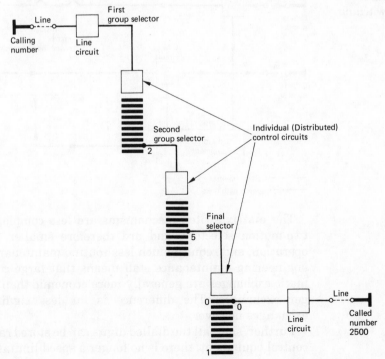

telephone instrument. The dial was designed to transmit each digit as an appropriate number of line-disconnect pulses at a rate of 10 per second, with a minimum time pause between successive digits to ensure satisfactory switching to one digit before the next one arrived.

The individual switching mechanism and associated relays and magnets required for each selector is large, heavy and expensive. It is noisy in operation, and requires a great deal of routine maintenance and adjustments to keep it operating satisfactorily. Automatic exchanges employing individual (distributed) control are therefore physically large, very noisy, and inefficient in operation because a large engineering maintenance staff is needed.

Common Control

Crossbar and **reed-relay automatic exchanges**, which employ matrix-type switches described earlier in the chapter, use a different method of switching control. When a call is initiated, the caller is connected to a *common control equipment* which accepts and stores all the dialled digits. It then establishes if the called number is free and then searches rapidly for a free route through the matrix switching network to connect the caller to the called number.

Once the call is set up, taking a very short time, the common control equipment is released to deal with other calls, but the connection between the two telephones through the matrix switching network is maintained until the conversation is finished. This is illustrated simply in Fig. 8.20.

Fig. 8.20 Principle of common-control exchange switching

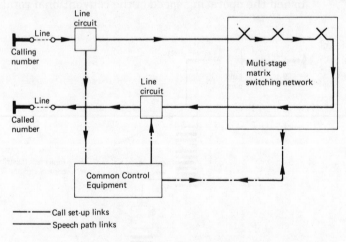

The matrix switch mechanisms are less complex then the Strowger two-motion selectors, and are therefore smaller, cheaper, quieter in operation, and require much less routine maintenance. The reduction in engineering maintenance staff means that large common control automatic exchanges are generally more economic than similar-sized Strowger exchanges. The difference is far less significant with smaller exchanges however.

Further, since all the dialled digits can be stored rapidly by the common control equipment, there is no longer a speed limitation, and the conventional rotating dial can be replaced by a push-button device, making called number selection easier and perhaps less prone to error.

Stored Program Control

The earliest common control equipment was completely electro-mechanical, using the same type of electro-magnet relays and electro-magnets employed in the Strowger two-motion selectors. The next stage was the use of early electronic devices such as thermionic and cold-cathode valves, but progress towards a new era of electronic common control equipment did not become significantly practical or economic until the advent of reliable transistors and printed circuit techniques.

At the same time, rapid progress was being made in the development of computers for business and commerce, and it was inevitable that efforts were directed to the possibility of controlling exchange switching equipment by computers. The unique problem was the fact that such control had to be reliable and satisfactory in real time, whereas most business and commercial applications could tolerate some degree of waiting time. Computerized common control equipment was first introduced to operate conventional crossbar and reed-relay matrix switches which provided space-division metallic contact routes for speech signals in conventional analog form. This arrangement was called *stored program control* (SPC) because it controlled the switching operations by a program stored in a memory that was capable of rapid alteration.

With the introduction of pcm systems providing digital transmission of speech signals by time-division techniques, it was inevitable that thoughts turned to the possibility of also using time-division for switching instead of space-division, leading eventually to computer-controlled digital exchanges. This has seen the emergence of exchanges with complete computer control not only of speech signals and switching functions, but also of administrative and management functions such as automatic billing, and the addition of many improved facilities for the telephone user. Such a system has been developed in the UK by British Telecom and industry under the intriguing name of System X.

9 Introduction to Radar and Navigational Aids

Principles

When a radio wave strikes an object or target, some of the energy is reflected or re-radiated by the target back towards the transmitting aerial, so that the presence of the target is *detected*. If the transmitted radio wave is in the form of pulses, then by measuring the time delay between the transmitted pulse and the received pulse or echo, the distance of the detected object from the transmitting aerial can be determined, since the velocity of propagation of radio waves in the atmosphere is constant at the speed of light (3×10^8 metres per second), and distance = velocity $\times$ time.

By using *highly directional* transmitting aerials that send a radio wave along a very narrow beam, it is possible also to identify the bearing or direction of the distant target relative to the transmitter.

Thus, we get a system of Radio Detection And Ranging, which prompted the introduction of the word **radar**. The first important point to make is that this system can only *provide* information, and does not carry information, as has been the case in all telecommunication systems previously considered.

The radar system briefly described above is known as a **pulsed primary radar system**. The radar transmitting aerial, as stated, is highly directional, and there are two main directions to be considered.

The first is the *horizontal* direction relative to the aerial. This is called the AZIMUTH, and the *beamwidth* of the aerial in this plane is very narrow – just a few degrees. The second is the direction upwards relative to the horizontal, that is the *vertical* direction or ELEVATION. This can be quite broad beam to accept reflected or echo signals from a wide range of vertical angles.

If this aerial is physically turned through 90° so that the original horizontal and vertical planes are interchanged, it will then give a narrow elevation beamwidth, with a broad azimuth beamwidth.

Another possibility is an aerial with narrow azimuth beamwidth and multiple narrow elevation beams, which can give a three-dimensional coverage of range, bearing and height, if the aerial is arranged to rotate or scan, on its own axis, in the horizontal plane. (See Fig. 9.1.)

Fig. 9.1 Two methods of narrow-beam scanning

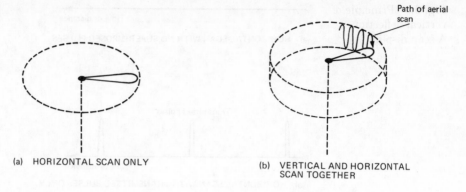

(a) HORIZONTAL SCAN ONLY

(b) VERTICAL AND HORIZONTAL SCAN TOGETHER

Fig. 9.2 Principle of measuring the range or distance

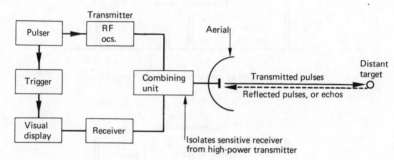

Measurement of Range or Distance

The principle of measuring the *range* or distance of a target is shown in Fig. 9.2.

The transmitter produces **pulses** of radio waves in the frequency range 150 MHz to 30 000 MHz (or 150 MHz to 30 GHz). The *duration* of the pulses will generally be between 0.25 and 50 microseconds (μs). The single aerial is usually used for transmitting and receiving pulses, and the *combining unit* shown in Fig. 9.2 isolates the sensitive receiver from the high-power transmitter pulses, and then switches the aerial to the receiver during the intervals between transmitter pulses. The *time interval* between pulses depends on the maximum distance at which the radar system is to be effective.

The transmitted and received pulses are displayed on a cathode-ray tube with a horizontal scan to give a *visual* indication of the time delay. The scan of the CRT display is synchronized with the transmitted pulses, and the transmitted and received pulses deflect the scan vertically to give a visual indication of each pulse.

Because of the constant velocity of propagation of the radio pulses, the time delay can be directly calibrated in distance. The time for the echo pulse to return to the receiver gives *total* distance out to the target and back again. This is illustrated in Fig. 9.3 and is called an A-scope or

Fig. 9.3 Principle of
vertical deflection
A-scan display

Time or distance

(a) HORIZONTAL SCAN WITH NO SUPERIMPOSED PULSES

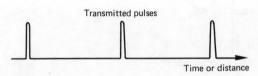

Transmitted pulses

Time or distance

(b) HORIZONTAL SCAN WITH TRANSMITTED PULSES ONLY

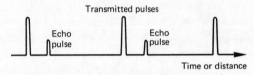

Transmitted pulses

Echo pulse

Echo pulse

Time or distance

(c) HORIZONTAL SCAN WITH CLOSE TARGET

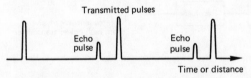

Transmitted pulses

Echo pulse

Echo pulse

Time or distance

(d) HORIZONTAL SCAN WITH DISTANT TARGET

A-scan display. Because the pulses deflect the scan on the CRT vertically
from its normal horizontal path, this type of visual indication is called
deflection modulation.

The transmitted pulses are actually short pulses of a high-frequency
carrier, as shown in Fig. 9.4. For example, if the radio frequency is 1 GHz
(1000 MHz), the time for one cycle (periodic time) is

$$T = \frac{1}{f} = \frac{1}{1000 \times 10^6} \text{ sec} = \frac{1}{1000} \mu s$$

So, if the length of the pulse is 1 μs, there will be 1000 cycles of carrier
frequency in each pulse. If the time interval between pulses is 1000 μs,
then the maximum effective range will be

$$\text{Velocity} \times \text{time} = (300 \times 10^6 \text{m/s} \times 1000 \times 10^{-6}\text{s}) \div 2$$
$$= 150 \text{ km} = 93.75 \text{ mile}$$

Note: the calculated time must be divided by 2 as shown because the pulse
actually travels a distance of 300 km, to the target and back.

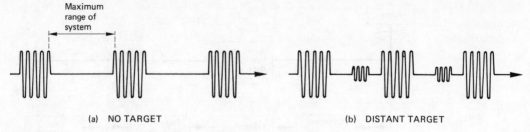

(a) NO TARGET   (b) DISTANT TARGET

Fig. 9.4 Actual form
of pulses

Measurement of Direction or Bearing

Having described a method of obtaining the distance or range of a target, we will now consider how to obtain direction or *bearing* of the target as well.

The highly directional aerial is rotated so that it searches over the whole of the horizontal plane during each revolution, which usually takes about $\frac{2}{3}$ second. As the aerial rotates, its bearing or direction relative to *north* for *azimuth* (and also relative to *horizontal* for *elevation* in some systems) is converted into either an analog or a digital electrical signal to apply to the visual display system.

The azimuth or horizontal bearing relative to north is displayed by a moving electron spot that traces out a narrow rotating beam on a circular cathode ray tube which has a long afterglow. The rotation of the trace is synchronized with the aerial rotation. The simple principle of this arrangement, which is called a **plan position indicator** (PPI), is illustrated in Fig. 9.5.

Fig. 9.5 Principle of
Plan Position Indicator
intensity display of
target bearing

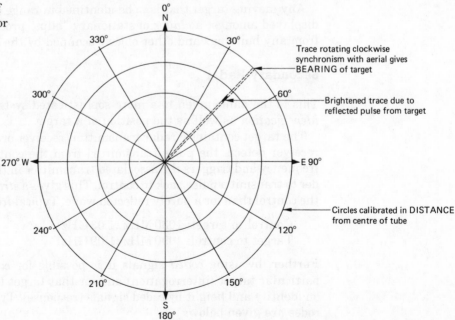

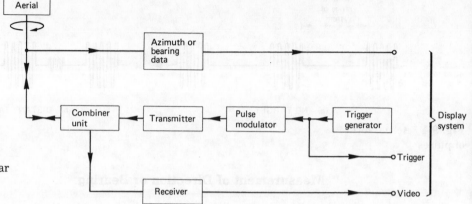

Fig. 9.6 Block diagram of simple pulsed primary radar with PPI display of target bearing

Any reflected or echo signal received from a target *brightens* the spot in order to give a visual indication of the range (by the distance along the scan from the centre) and the bearing (from a calibrated compass scale) in the horizontal plane. A block diagram of this simple pulsed primary radar system is given in Fig. 9.6. Since the receiver echo pulses brighten the trace to give visual indication of distance and bearing, this system is called **intensity modulation**.

For a pulsed primary radar system fitted on board a ship, the "north" position corresponds to the bow of the ship. In the case of a ground pulsed primary radar system tracking aircraft that are close together, it is necessary to request individual aircraft by radio link to carry out a procedural turn to left or right in order to identify each one. This is time-consuming, expensive in fuel, and increases the load on the radio communication equipment at the control point.

Any moving target that is to be identified in range and bearing will be displayed amongst a *clutter* of stationary "blips" produced by reflections from any buildings and other objects scanned by the rotating aerial.

Secondary Radar

This is the name given to a more sophisticated system which gives the *identification* as well as the position of a target.

The target contains a radar transmitter/receiver or **transponder**. The receiver detects the pulse transmitted from the control at a particular frequency and triggers the associated transmitter in the target transponder to transmit a pulse back to control. This gives a *stronger* echo signal at the control than for a purely reflected wave. Typical frequencies would be

Control to target: 1030 MHz (1.03 GHz)
Target to control: 1090 MHz (1.09 GHz)

Further, by using *coded* signals it is possible for control to address a particular target (interrogation) and for that target to give information on identity and height by coded signals (response). Typical interrogation codes are given below:

Interrogation Code	Pulse Spacing (microseconds μs)	Use
1	3	Military identity
2	5	Military identity
3/A	8	Joint military/ civil identity
B	17	Civil identity
C	21	Altitude
D	25	Unassigned

Since the control display will only respond to predetermined re-transmitted signals from a definite target, the problem of clutter is greatly reduced, since unwanted stationary objects will *not* re-radiate a strong signal at an appropriate frequency.

The main advantages of secondary radar over primary radar systems can be summarized as

1 Larger echo signal at the control receiver.
2 Identification of target as well as its position.
3 Targets addressed only when needed by control.
4 Variety of information possible from targets.
5 'Clutter' greatly reduced, giving a moving-target indicator system (MTI).

Radar can be employed in several ways as a navigational aid for civil use and for several military uses.

CIVIL USES

1 Since mountains, plains, cities, rivers, oceans, etc. all reflect radio waves to different extents, radar equipment in aircraft can provide information about the ground below despite poor visibility or darkness.
2 Radar equipment on a ship can similarly give information on the location of other ships, marker buoys, land, etc. despite poor visibility or darkness.

Also, radar systems can be used to

3 Give aircraft and ships information as to their exact position at any moment.
4 Assist aircraft in landing during poor visibility or darkness.
5 Enable aircraft to measure height above ground.
6 Enable air space above an airport to be monitored to control take-off and landing of aircraft.
7 Check speed of vehicles on motorway.

MILITARY USES

1 Aiming ground-based guns at ships and aircraft.
2 Locating ground targets for bomber aircraft through clouds or at night.
3 Locating moving targets for fighter aircraft at night.

4 Directing guided missiles from ground, ships or aircraft.
5 Tracking enemy missiles or aircraft to give early warning systems.
6 Searching for submarines.

Radio Navigational Systems

Radio waves have long been used as an aid to navigation.
Some of the many different systems will now be discussed.

Visual–Aural Range (VAR) This is a ground-based system which enables an aircraft to determine its position relative to the ground transmitter.

Four separate signals are transmitted at the same radio frequency from directional aerials. Two of the transmissions are in the form of Morse Code signals, one representing letter N (dash dot) and the other representing letter A (dot dash). The other two carriers are modulated by 90 Hz and 150 Hz tones respectively.

In an aircraft, the receiver demodulates the four signals, converting the N and A signals into *audible* tones for the pilot, and converting the 90 and 150 Hz tones into d.c. signals which indicate blue or yellow respectively on an instrument. The waveform and directivity of the four transmitted signals are illustrated in Fig. 9.7.

There is a range area around the four transmitting aerials in which the signals are workable, and the range area is divided into four quadrants by the directional aerials. This is illustrated in Fig. 9.8. Each quadrant contains a particular *pair* of signals, which enables the pilot in an aircraft to determine which quadrant he is in by the two demodulated signals received. For example.

Quadrant 1 – yellow and N
Quadrant 2 – yellow and A
Quadrant 3 – blue and A
Quadrant 4 – blue and N

Along the line separating the Blue and Yellow quadrants, the d.c. outputs produced by the demodulation of the 90 Hz and 150 Hz tones cancel so that the receiver instruments indicates neither blue nor yellow.

Along the line separating the N and A quadrants, the receiver output gives a continuous audible signal, instead of either N or A notes.

Visual Omnirange (VOR) This system enables the pilot of an aircraft to read an exact bearing on a single instrument. Two signals are radiated from the ground transmitter. One signal, called the *reference* signal, is obtained by a 30 Hz tone frequency-modulating a 9960 Hz sub-carrier, and the resulting fm wave then amplitude-modulates a radio frequency carrier. This signal is radiated in all directions from the transmitter.

The second radiated signal is obtained by a 30 Hz tone amplitude-modulating the carrier, whose phase is varied according to the *direction* of radiation from the transmitter. The two radiated signals are *in phase* for a direction due *south* of the transmitter, and the phase difference between

Fig. 9.7 The four directional signals radiated in VAR

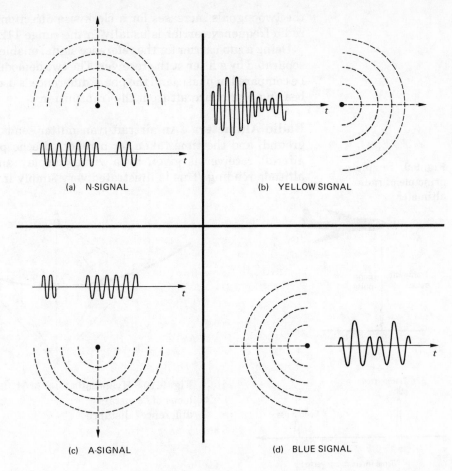

(a) N-SIGNAL

(b) YELLOW SIGNAL

(c) A-SIGNAL

(d) BLUE SIGNAL

Fig. 9.8 Transmitted wave from the four VAR aerials

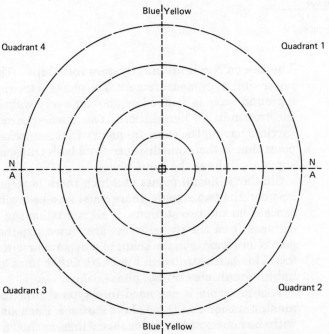

the two signals increases for a clockwise direction from due south. The radio frequency carrier is usually in the range 112 MHz to 118 MHz.

Using a sub-carrier for the reference signal enables the two signals to be separated by a filter at the receiver. The two demodulated 30 Hz tones can be compared in phase, and the phase difference is used to indicate an exact bearing of the aircraft from the transmitter.

Radio Altimeters An aircraft transmitter sends a signal down to the ground, and the time taken for a reflected echo pulse to return to the aircraft receiver is given by a visual display and converted into an altitude reading. This is illustrated very simply in Fig. 9.9.

Fig. 9.9 Simple principle of radio altimeter

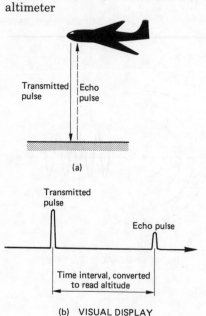

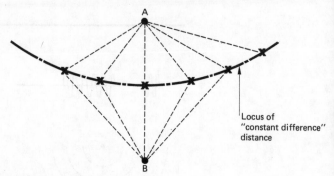

Fig. 9.10 Hyperbolic locus of "constant-difference" distance

The Decca Navigational System for Ships This system is based on a *phase-difference* measurement. The phase of the radio-frequency carrier of a ground wave is delayed by 360° for every wavelength of distance from the transmitting aerial. So, if two fixed transmitting stations radiate carriers having the same frequency that are locked in phase, then at one point that is the same distance from both transmitters, the two received signals will have the same phase.

Similarly, for all points at which there is a constant phase difference between the two signals, there must also be a constant difference of distance from the two stations. If all points having a constant "difference" distance from the two stations are joined together, the line joining the points (or locus) is in the shape of a hyperbola with the two stations as the foci. This is illustrated in Fig. 9.10. Other lines (or loci) can be added to indicate multiples of 360° phase delay.

This principle is extended in a typical practical system by having a **master** station and three **slave** stations. Each slave station forms a pair with the master so that three sets of lines as shown in Fig. 9.10 intersect to

Fig. 9.11 Decca
Navigator system with
master (A) and three
slaves (B, C, D)

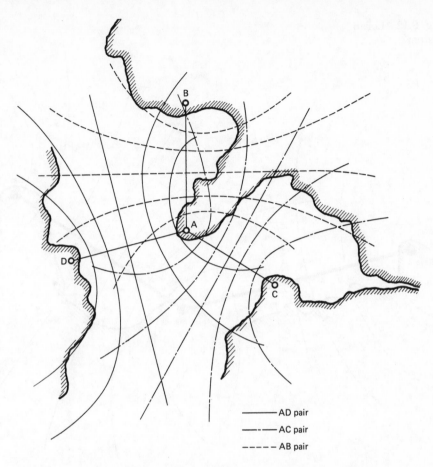

AD pair
—·—·— AC pair
— — — — AB pair

give *position fixing* information through the whole 360° of azimuth (horizontal direction from transmitter).

The space between the hyperbolic lines which correspond to adjacent multiples of 360° phase shift is called a *lane*. Several installations as illustrated in Fig. 9.11 are provided at different points along various coastlines in an area (e.g. Western Europe) to give a Decca Navigator Chain.

Loran System The LOng RAnge Navigational aid system was developed in the USA during the last war as a long-distance navigational aid for aircraft and ships. This system is based on a *pulse arrival-time difference* measurement as opposed to the phase difference measurement of the Decca system.

Fixed transmitting stations are arranged in pairs. One station of a pair transmits a pulse at a given frequency, and this pulse is received by the other station of the pair. The second station then transmits a pulse on the same frequency as the first station after a definite time delay.

There will be a number of points at which the two transmitted pulses will arrive with a constant time difference, representing also of course a constant path-length difference. As for the Decca system, these "equal

Fig. 9.12 Loran
system

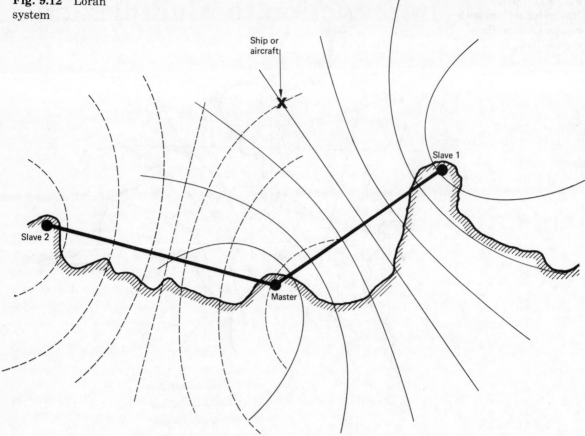

difference" points will be on a line or locus that has the shape of a hyperbola, with the two transmitting stations as foci. The pulses are transmitted at a rate of approximately 30 times per second. "Position lines"
can be set up by measuring differences in pulse-arrival times.

The measurement is achieved by a visual display on a CRT, with time
bases locked and provided with timing marker "blips" derived from a
100 kHz crystal-controlled oscillator in the receiver.

A second pair of stations is then established to provide another set of
"position lines" by transmitting pulses on the same frequency as the first
pair. A slightly different pulse recurrence rate is used, however, so that
the visual display of the second pair of pulses slowly drifts across the time
bases locked to the first pair. Thus the two displays can be distinguished.

Generally, a common control station, or Master, will be used for two
pairs, using two slave stations, with a station separation of 320 to
480 kilometers (200 to 300 miles). The radiated frequency is around
2 MHz. Fig. 9.12 illustrates the principle of the Loran system.

10 Introduction to Multiplexing

Introduction

Chapter 2 (page 13) introduced the concept of multiplexing as using a transmission line for the simultaneous handling of several telephone speech channels. This is necessary in a rapidly expanding national network, particularly for long-distance trunk calls, because the number and cost of trunk lines needed to handle the telephone calls on a "one conversation per line" basis would be prohibitive.

Any transmission line has a maximum frequency bandwidth available for any given period of time. Multiplexing can be achieved by either frequency-division or time-division techniques.

With **frequency-division multiplex** (fdm), each speech channel is allocated a unique portion of the available line bandwidth, and has exclusive use of this allocated bandwidth all the time, as illustrated in Fig. 10.1a.

With **time-division multiplex** (tdm), each channel is allocated the whole of the line bandwidth for specific regular periods of time, the time-slots obviously depending on the number of speech channels sharing the line, as illustrated in Fig. 10.1b. In each case, guard bands separate adjacent channels as protection against mutual interference.

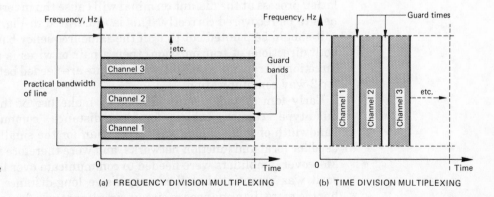

Fig. 10.1 Simple illustration of multiplexing

(a) FREQUENCY DIVISION MULTIPLEXING

(b) TIME DIVISION MULTIPLEXING

Frequency-division Multiplex

As previously described in Chapter 2 (page 12), each telephone speech channel is limited to a bandwidth of 300–3400 Hz (commercial speech), and is used to amplitude-modulate an allocated carrier frequency to produce upper and lower sidebands (see Fig. 2.7). The upper sideband and the carrier frequencies can be suppressed by the modulator and a suitable bandpass filter, so that the original speech channel information appears to have been shifted to a particular position in the frequency spectrum of the line, depending on the actual carrier frequency, as illustrated in Fig. 10.2.

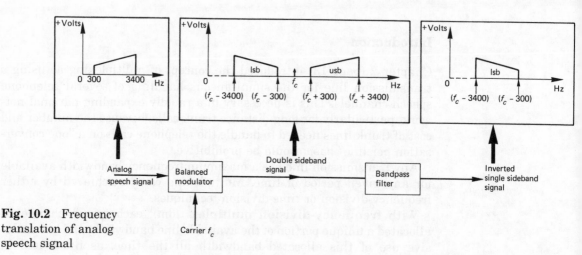

Fig. 10.2 Frequency translation of analog speech signal

So, for multi-channel operation, if other speech channels modulate different carrier frequencies spaced at 4 kHz intervals, then all the original speech channels can be assembled within the available bandwidth of the line for simultaneous transmission, as illustrated in Fig. 10.3.

In this simple arrangement the speech channel frequencies have been inverted in the modulation process; that is, the highest frequency of each channel is the lowest in the frequency spectrum of the line. The demodulation process at the distant terminal will cause the inverted speech channels to be recovered correctly. This is illustrated in Fig. 10.4.

It should be pointed out that, if the same frequency bands are used for both directions of transmission, then a pair of wires is needed for each direction of transmission, so 4-wire circuits are needed between terminals for 2-way communication.

Early fdm systems were designed to make use of the bandwidth of quad-type carrier cables for long-distance communication. The bandwidth of these cables was larger than for the smaller-gauge cables used for local and junction networks, and were therefore more expensive. Moreover, amplifiers were needed to communicate over longer distances, so it was essential that these expensive long-distance circuits should handle more than one speech circuit simultaneously for economic reasons.

Fig. 10.3 Assembly of multichannel line signal

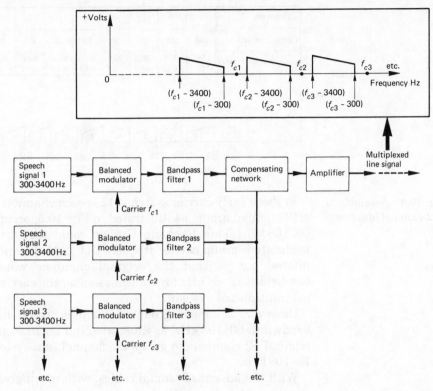

Fig. 10.4 Demodulation of fdm line signal

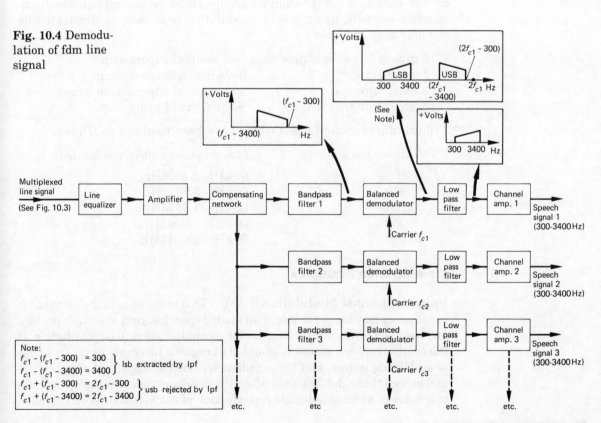

Note:
$$f_{c1} - (f_{c1} - 300) = 300$$
$$f_{c1} - (f_{c1} - 3400) = 3400 \Big\}\ \text{lsb extracted by lpf}$$
$$f_{c1} + (f_{c1} - 300) = 2f_{c1} - 300$$
$$f_{c1} + (f_{c1} - 3400) = 2f_{c1} - 3400 \Big\}\ \text{usb rejected by lpf}$$

Channel no.	1	2	3	4	5	6	7	8	9	10	11	12
Carrier freq. (kHz)	108	104	100	96	92	88	84	80	76	72	68	64
Lower sideband (kHz)	104.6 -107.7	100.6 -103.7	96.6 -99.7	92.6 -95.7	88.6 -91.7	84.6 -87.7	80.6 -83.7	76.6 -79.7	72.6 -75.7	68.6 -71.7	64.6 -67.7	60.6 -63.7

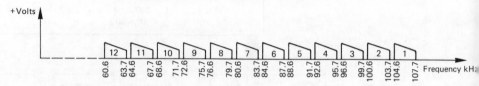

Fig. 10.5 Assembly of a 12-channel fdm group

In these early carrier systems, 12 speech channels were assembled into a 12-channel group, as illustrated in Fig. 10.5, occupying a bandwidth of 60–108 kHz. This 12-channel group could then be used to amplitude-modulate a group carrier frequency of 120 kHz, and the lower sideband filtered out so that the original channels were assembled in the bandwidth 12–60 kHz for transmission on the carrier cable as a 12-channel multiplexed system.

However, a second group of channels could be assembled in the bandwidth 60–108 kHz, as illustrated in Fig. 10.5, and then added to the original 12 channels to give a 24-channel fdm system in the bandwidth 12–108 kHz.

With the advent of coaxial cables, with much greater bandwidth than carrier cables, many 12-channel groups could be assembled into high-capacity systems, using several modulation processes, as illustrated by the following examples:

5 groups	could produce a	60-channel supergroup.
5 supergroups		300-channel mastergroup.
3 mastergroups		900-channel supermaster group.
15 supergroups		900-channel hypergroup.

Typical fdm coaxial systems which have been used are as follows:

Number of channels	Line frequency spectrum needed
600	60 kHz–2.46 MHz
960	60 kHz–3.90 MHz
1 200	60 kHz–4.86 MHz
3 000	60 kHz–12.06 MHz
10 000	60 kHz–40.06 MHz

Time-division Multiplex

Pulse Amplitude Modulation (PAM) This makes use of the fact that it is not necessary for the *whole* of an analog speech signal to be transmitted and received continuously in order for communication to be intelligible. If the original speech signal is sampled at regular intervals to give a series of amplitude pulses, and these pulses are then transmitted along a line and passed through a low-pass filter, the resultant received signal appears to a listener as an intelligible reproduction of the original speech signals.

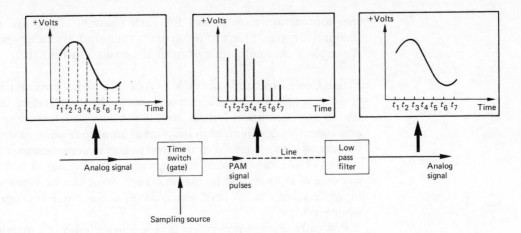

Fig. 10.6 Simple illustration of pulse amplitude modulation (pam) transmission

For this to be acceptable, the analog signal must be sampled at approximately double the highest frequency contained in the analog signal. For commercial speech, the highest frequency is limited to 3.4 kHz, so a sampling rate of 8 kHz is used. This simple process is called pulse amplitude modulation, with pam pulses being transmitted along the line, as illustrated in Fig. 10.6.

Now, if other analog speech signals are sampled at the same rate but at different times, all the resulting pam pulses can be interleaved in unique time-slots for transmission along the same line as a multiplexed signal. At the receiving end of the line, the multiplexed line pam pulses must be sampled at the same times as those at the sending end in order to extract

Fig. 10.7 Simple 2-channel pcm system

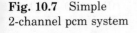

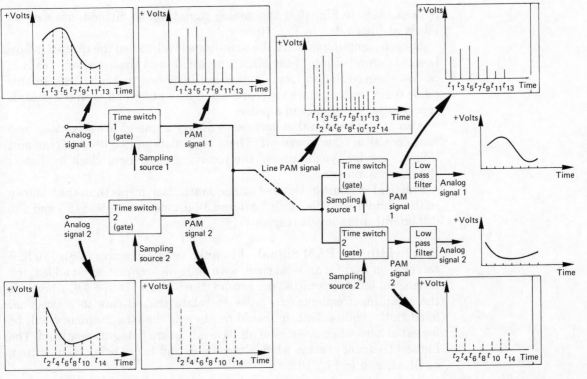

the appropriate pam pulses for each channel. They are then passed through low-pass filters to reconstruct the original analog speech signals. A simple 2-channel pam system is illustrated in Fig. 10.7.

Pulse Code Modulation (PCM)   As explained, a pam signal has pulses of varying amplitudes, and can be distorted by interference and noise on a line of any significant length, so it is not very satisfactory or reliable. To overcome this problem, each individual amplitude pulse of the pam signal is encoded into a series of equal-level pulses in accordance with the binary system. The total range of voltages of an analog signal is divided into a number of equi-spaced levels, and each level can be represented by its binary equivalent. (A brief explanation of the binary system is given in Chapter 11.)

Basically, the binary system is based on an index of 2 instead of 10 as in the denary (decimal) system. The two possible states (pulse or no pulse) are represented by 1 and 0 respectively, and different numbers of binary digits (bits) can be used, according to requirements.

Now, for n bits, 2^n possible values can be represented, illustrated simply as follows:

Number of bits	Possible states	Denary numbers							
		0	1	2	3	4	5	6	7
$n = 1$	$2^1 = 2$	0	1						
$n = 2$	$2^2 = 4$	00	01	10	11				
$n = 3$	$2^3 = 8$	000	001	010	011	100	101	110	111

Binary equivalents

So, returning to Fig. 10.6, the analog signal can be divided into 8 equal levels as illustrated in Fig. 10.8a.

At each sampling instant, the amplitude level (called the **quantization level**) is encoded into 3-bit binary notation, and the resultant train of pulse-code-modulation (pcm) pulses is illustrated in Fig. 10.8b. In this pulse train, each 1 (mark) is represented by a pulse and each 0 (space) is represented by absence of a pulse.

Also, it will be seen that successive binary numbers are separated by a space equal in time to one bit. These time-slots can be used to transmit information for synchronizing the receiving equipment clock to that at the transmitting end.

Practical systems in fact use more quantization levels than the 8 shown in this simple example, with 7-bit and 8-bit codes giving 128 (2^7) and 256 (2^8) quantization levels respectively.

Bandwidth of a PAM Signal   From the simple example of Fig. 10.8, it can be seen that for an n-bit code, with one *synchronization bit* added, the number of bits transmitted per second (the bit-rate) is $(n + 1)f$, where f is the sampling frequency. In order to relate the bit-rate to a practical bandwidth requirement, it should be obvious that the frequency will be lowest at zero when a series of all 0s or all 1s are being transmitted. The highest frequency occurs when alternate 0s and 1s are being transmitted, as illustrated in Fig. 10.9.

Fig. 10.8 Illustration of quantization and coding

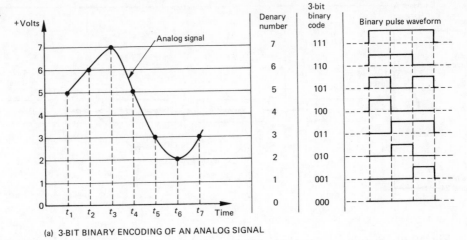

(a) 3-BIT BINARY ENCODING OF AN ANALOG SIGNAL

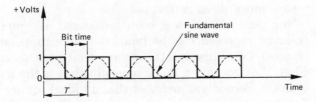

(b) PCM PULSE TRAIN REPRESENTING ANALOG SIGNAL IN (a)

Fig. 10.9 PCM waveform having highest frequency

The periodic time T of the pcm waveform of Fig. 10.9 is clearly double that of a single bit, so the frequency of this waveform is half the bit-rate, since frequency is the reciprocal of time.

The bandwidth required for any pcm waveform is therefore the difference between the highest and lowest frequencies, that is

$$\left(\frac{\text{Bit-rate}}{2} - 0\right)\text{Hz} \quad \text{or simply} \quad \frac{\text{Bit-rate}}{2}\text{Hz}$$

Now, as previously stated, the speech analog signal is sampled 8000 times a second, or at 8 kHz, so that each time-slot is 1/8000 seconds, or 125 μsec.

In practice, each binary-coded sample level can be transmitted in a few microseconds, so there is a lot of unoccupied time in each 125 μsec slot. This time can be used to transmit pcm pulses obtained from other speech

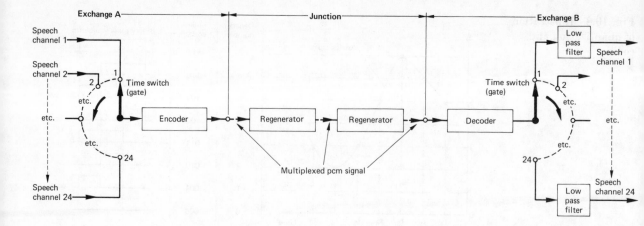

Fig. 10.10 Simple illustration of one-way pcm system

analog signals, and so a multi-channel multiplexed system can be arranged on a single line. Again, as for fdm, a 4-wire circuit is needed to give two-way communication between terminals. The principle of such a system is illustrated in Fig. 10.10.

Line distortion The pcm pulses are attenuated, delayed and distorted by the electrical characteristics of the line. If the receiving equipment can determine whether a pulse is present or not amongst the interference and noise, then no errors are introduced. For this to be possible over relatively long lines, **pulse regenerators** are inserted at regular intervals along the line. The length of each section of line is such that the received pcm pulses can be detected satisfactorily, and then regenerated and transmitted onwards as perfect pulses.

The pcm pulses have been considered as *unipolar*, that is positive polarity representing 1s (marks) and zero polarity representing 0s (spaces). Transmission of unipolar pulse trains has certain disadvantages. First, they contain a d.c. component that is the mean of all positive 1s (marks). Secondly, remember that an ideal square waveform contains a fundamental sinusoidal frequency and all the odd harmonics to infinity, and the lower-frequency harmonics will have large amplitudes. Furthermore, there can be long periods of consecutive 1s or consecutive 0s.

These factors increase the complexity of the regenerators placed at regular intervals along the line. To simplify the design of the regenerators, the pcm unipolar pulse trains are applied to a process of *alternate digit inversion* (ADI) followed by a process of *alternate mark inversion* (AMI), as illustrated in Fig. 10.11. The resulting pcm signal can now be transmitted and regenerated as required at regular intervals with minimum complexity and cost.

However, although AMI ensures that alternate 1s have opposite polarity, there is still a major disadvantage in the fact that a train of successive 0s can give a long time with no signal on the line. The pulse regenerators use the marks to obtain timing information, so their function can be seriously affected by a long train of spaces.

This problem is overcome by the use of High Density Bipolar 3 (HDB3) techniques, where a maximum of 3 successive 0s is automatically followed

Fig. 10.11 Simple illustration of ADI and AMI

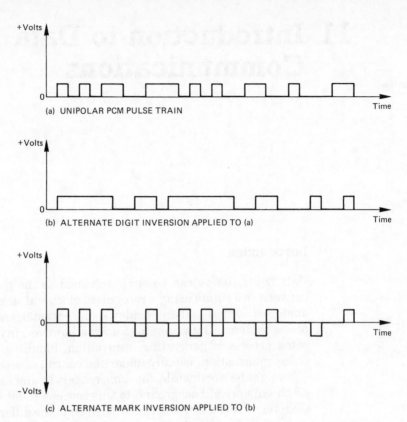

(a) UNIPOLAR PCM PULSE TRAIN

(b) ALTERNATE DIGIT INVERSION APPLIED TO (a)

(c) ALTERNATE MARK INVERSION APPLIED TO (b)

by a 1 (called a violation mark). The receiving equipment recognizes that this is a violation of AMI rules, and automatically interprets it as another 0.

Practical PCM Systems The first practical use of pcm multiplex systems was made in the junction network interconnecting exchanges in a fairly restricted geographical area (as opposed to trunk lines as for fdm systems). The junction cables in use on a "one-call-per-line" basis had loading coils inserted at regular intervals (typically 2000 yards) to compensate for the increasing attenuation at the higher frequencies in the commercial speech range (300–3400 Hz).

It was found that this spacing was suitable for the pcm regenerators to replace the loading coils, and thus give multi-channel operation instead of single-channel operation on each junction.

Typical systems in use by British Telecom are:

Channels	Sampling rate	Bit-code	Bit-rate (kbit/s)	Bandwidth (kHz)
24	8 kHz	7 (plus 1 sync)	$8 \times 8 \times 24 = 1536$	768
30 (plus 2 channels for signalling and synchronization)	8 kHz	8	$8 \times 8 \times 32 = 2048$	1024

11 Introduction to Data Communications

Introduction

Data *transmission* can be simply defined as the passing of information between two points using a recognized electrical signal code, instead of by analog speech signals as achieved by national and international telephone systems. Data *communications*, however, involves a more complicated process of generating, controlling, handling and checking of the coded information, usually under the control of a computer.

It might be worthwhile, for some readers at any rate, to consider first of all the historical background to this concept of data communications. In Chapter 1 a brief introduction was given to the difference between analog and coded signals, with a very early code being described and illustrated in Fig. 1.4, called the Morse Code. This enabled two persons to communicate with each other beyond the normal distance of human voice and hearing by means of *visible* signals, and later beyond the range of human vision by line or wireless transmission which produced an *audible* coded signal at the destination. In order to communicate in this way, both sender and receiver needed to know the Morse Code as recognised internationally, and to be able to operate at transmitting speeds that would make messages as economical as possible. The receiver decoded the message as it arrived and produced a handwritten copy, which gave rise to the term telegraphy, or "writing at a distance". It was a natural progression, of course, that thoughts and efforts were soon directed at replacing humans by machines, as has always been the case as civilization has continually developed.

One reason was the fact that machines might increase speed of transmission, thus increasing the effective use of telegraph circuits. Another reason was the ability to transmit to unmanned machines when normal working hours do not coincide, as for example on opposite sides of the world.

The nature of the Morse Code however presented a problem for machine operation. The number of basic signal elements for individual characters varied greatly, so the time taken to send different characters varied greatly also. For example, a single "dot" represented letter E, and five "dashes" represented figure 0, with various intermediate combinations of 1, 2, 3, 4 and 5-element codes representing other letters and figures.

This presented no problem for human operation, but it would have been very difficult and expensive to design a machine capable of identifying characters with such a variation in signal elements and transmission times. As a result, several possible codes were proposed, and eventually one was agreed internationally in 1932 based on the work of Murray, Baudot and others.

This code used consecutive equal-duration pulses of negative and positive 80 volts, called **marks and spaces** respectively, in groups of five and was eventually designated International Alphabet No. 2 (IA2). To enable the receiving machine to detect the beginning and end of each character code, each group of five marks/spaces was preceded by a start signal and followed by a stop signal. The start signal was one element of space polarity and the stop signal was $1\frac{1}{2}$ elements of mark polarity. It is interesting to note that this start/stop technique is still used in some modern data communication systems.

Using five-element combinations of only two possible voltage polarities enables 32 (2^5) different characters to be represented. This was inadequate for 26 letters, 10 figures and other functional characters such as full stop, comma, plus, minus, equals, question mark, brackets, inverted commas, etc. that are necessary for normal language communication. So, the machine was designed to include a mechanical letter/figure shift facility to allow each five-element code combination to represent *either* of two

Fig. 11.1 Typical characters of IA2 code

"Letters" character	Alternative "figures" character	5-unit code
A	–	MMSSS
B	?	MSSMM
E	3	MSSSS
R	4	SMSMS
Y	6	MSMSM

Fig. 11.2 Electrical signal representing letter Y in IA2 code

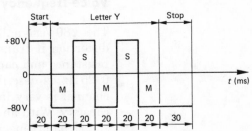

possible characters, as required. The arrangement extended the number of possible characters to a maximum of 64.

By way of illustration, Fig. 11.1 shows some typical 5-unit character codes (ignoring the start and stop elements) and Fig. 11.2 shows the complete electrical signal for letter Y. Since the individual signal elements of any character are sent one after another, this is known as **serial transmission** of the information elements, a term which is also used for data transmission nowadays.

Early machines designed to use this five-unit code printed characters on a continuous paper tape which could be cut up and pasted on to a message form. This was typical of the machines used for public telegram services which no longer exist in the UK. Later machines introduced page-printing facilities like a typewriter for fairly obvious requirements. Both tape and page-printing machines were called *teleprinters* (or *teletypes*), and also like typewriters had a keyboard for selecting characters for transmission. When a character was selected by depressing a key on the keyboard at the sending end, the machine automatically generated the appropriate

$7\frac{1}{2}$-unit electrical coded signal which was pased to the line connected to the receiving machine. The machine recognized the signal and the mechanism automatically printed the appropriate character on a tape or a page.

Signalling Speed

The speed of signalling with the IA2 code was expressed in two ways. First, the reciprocal of the basic signalling element gives a signalling speed in **Bauds**. Again, we shall see later that this has a modern-day equivalent in data communications. For example, the basic signal element in Fig. 11.2 is seen to be of 20 ms duration, giving a signalling speed of $1/(20 \times 10^{-3})$ Bauds or 50 Bauds. Alternatively, for letter Y shown in Fig. 11.2 the 5-unit code consists of alternate marks and spaces, giving an effective alternating waveform having consecutive complete cycles of 40 ms duration. So, the fundamental sinusoidal frequency corresponding to letter Y is therefore $1/(40 \times 10^{-3})$ Hz or 25 Hz. This waveform represents the highest possible fundamental frequency generated by the 5-unit code. Note from Fig. 11.2 that letter R also generates the highest frequency, but the alternate marks and spaces are reversed in time. The lowest frequency occurs with five consecutive marks or five consecutive spaces, both of which produce a d.c. signal having zero frequency. So, the fundamental frequency range generated by the IA2 code is from 0 to 25 Hz.

Voice-frequency Telegraphy

The ± 80 V signal pulses produced by a teleprinter would suffer severe distortion if transmitted over long distances due to the effect of line resistance and capacitance. Further, with a working bandwidth of only 0–25 Hz as described above, the use of a line for only one teleprinter channel is very inefficient, since the line has a much wider bandwidth capacity. With an established telephone network having a fundamental speech band capacity of 300–3400 Hz for each channel, it was a logical step to consider using speech channels for teleprinter working. By using a number of teleprinter ± 80 V signals to amplitude-modulate different audio-frequency carrier frequencies spaced at intervals of 120 Hz, several teleprinter links (typically 18 to 30) could be transmitted simultaneously on a normal analog speech channel having a bandwidth of 300 to 3400 Hz. The bandwidth occupied by each teleprinter channel would be ± 25 Hz on the allocated carrier frequency. The simple principle of VF telegraphy is illustrated in Fig. 11.3. Again, the technique can be considered as the forerunner of the use of modems (*modulator/dem*odulator) in present-day data communications.

Development of Modern Data Communications

The growth of the application of computers has been very rapid in many fields since the advent of semiconductors, transistors and integrated solid-state components and circuitry. Prior to this, early computers had used electromagnetic relays and thermionic valves. These computers were largely used for complex mathematical problems, were large and

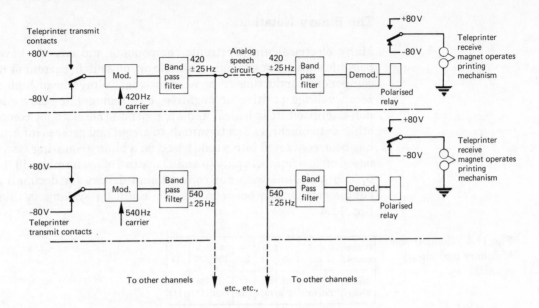

Fig. 11.3 Simple principles of VF telegraphy (unidirectional only)

cumbersome requiring large electrical power sources, and produced a great deal of heat. Generally, in most cases in commerce and industry, they were hidden away in a central computer room, and the tasks they performed were brought to the computer for processing on a punched paper tape, punched card, magnetic tape or keyboard printer for attention in due course.

The new generation of computers brought about a rapid increase in the number and types of task they could suitably handle in all areas of business, commerce and industry. As well as solving complex mathematical problems, computers became capable of high-capacity storage and retrieval of information data, and of handling routine functions such as invoicing, payrolls, records and statistics, etc., at a much faster rate than was possible by humans hitherto. It soon became clear that to make the most efficient use of expensive computer installations it was necessary to extend their availability beyond an isolated central location to intercommunicate with a number of different remote control positions or terminals, and also with other computers. In order to get information into and out of a computer it is necessary to use a suitable coded language. All information data normally recognized by humans in several different forms must be processed into a common code that the computer can accept. As previously stated, data can be processed by several peripheral means, for example paper tape, puncher/reader, punched card, keyboard printer, magnetic tape or disc, etc. Generally, the computer can function much more rapidly than a single input device, so it can deal with many devices simultaneously.

Data can be applied directly to the computer from various devices to be processed immediately in **real time**. This is known as **on-line** working. Alternatively, data from whatever source can be stored temporarily and applied to the computer later as appropriate, as described previously for early computer operation. This is known as **off-line** working, and does not occur in real time.

The Binary Notation

Many electrical and electronic components and circuits have just *two* possible states. For example, a simple on/off switch; current or no current; component conducting or non-conducting; voltage level high or low (or zero); voltage positive or negative; electromagnetic relay energized or non-energized. It is logical and appropriate therefore for computers and other data machines and terminals to accept and process information that has been converted into a code based on a binary counting system. This is based on *two* numbers only (0 and 1) instead of *ten* numbers (0, 1 , 2, 3, 4, 5, 6, 7, 8 and 9) as with our conventional denary (or decimal) system. A simple comparison between the two counting systems is illustrated in Fig. 11.4.

Fig. 11.4 Comparison of denary and binary counting

Denary	0	1	2	3	4	5	6	7
Binary	0	1	10	11	100	101	110	111

Denary	8	9	10	11	12	13	14	15
Binary	1000	1001	1010	1011	1100	1101	1110	1111

→ etc.

Now, in the denary system using a base of 10, we can see for example that number 4857 can be written as

$$4000 + 800 + 50 + 7$$
or $$(4 \times 10^3) + (8 \times 10^2) + (5 \times 10^1) + (7 \times 10^0)$$

(remember that any quantity to the power zero is 1).

In the same way, in the binary system using a base of 2, we can see that number 1111 can be written as

$$(1 \times 2^3) + (1 \times 2^2) + (1 \times 2^1) + (1 \times 2^0)$$
or $$8 + 4 + 2 + 1 = 15 \quad \text{(check in Fig. 11.4)}$$

Fig. 11.4 shows that
1 binary digit is needed to represent denary numbers 0 and 1.
2 binary digits are needed to represent denary numbers 2 and 3.
3 binary digits are needed to represent denary numbers 4, 5, 6 and 7.
4 binary digits are needed to represent denary numbers 8 to 15.
etc.
However, we can use 4 binary digits to represent denary numbers 0 to 15 (i.e. 16 numbers altogether) by placing zeros in the unused positions, resulting in the complete representation of all the denary base numbers 0 to 9, with 6 extra possible representations.

For example, denary 1 can be represented by binary 0001
denary 5 can be represented by binary 0101
denary 7 can be represented by binary 0111   etc.
So generally, for n binary digits we can represent 2^n denary values or combinations, so that for $n = 4$, $2^4 = 16$ possible numbers can be represented as previously seen above.

Each 0 or 1 is called a *binary digit*, which is suitably contracted and called a **bit**. The number of consecutive bits used or transmitted in one second represents the signalling speed in bits per second (bit/s).

A *group* of consecutive bits which form a particular number or character (i.e. 4 bits in the foregoing example, but other group sizes can be used) is called a **byte**. The bits contained in a byte can be used or transmitted consecutively (one after another) in *serial* form, as previously explained for the 5-unit IA2 code, or they can be used or transmitted simultaneously in *parallel* form.

We have already seen that for teleprinters using the IA2 code, each element or bit lasts 20 ms, giving a signalling speed of $1/(20 \times 10^{-3})$ Bauds, or 50 bit/s. Thus we have a similarity between the old teleprinter code and the method of specifying speed of transmission in the binary code.

Computer Codes

Any computer or data machine or terminal needs a suitable binary-coded language that covers all the possible alphanumeric characters (letters and numbers) plus other assorted characters (as previously described for a teleprinter) including a number of data link control functions known as the *protocol*.

One system, using 6-bit bytes, can give the full 36 alpha-numeric characters (letters and figures) plus other necessary information and control characters up to a maximum of 64 (2^6). This is called the Binary Coded Decimal (BCD) system. However, for a number of reasons, computer codes were developed with 7-bit or 8-bit bytes, as will be seen later.

The hexadecimal system It is worth mentioning at this point that an 8-bit byte, sometimes called an *octet*, can be represented more simply by the hexadecimal system (base 16). Any 8-bit byte can give 256 (2^8) different combinations of 0s and 1s ranging from 00000000 to 11111111. Each octet can be considered as two half-octets each having 4 consecutive bits. We have previously seen that there are 16 possible combinations of 0s and 1s using a 4-bit byte, and these 16 combinations can be more simply represented by a code called the hexadecimal code (hex) as illustrated in Fig. 11.5. So each half-octet of any 8-bit byte can be represented instead by the hex code.

Fig. 11.5 Comparison of denary, binary and hexadecimal counting

Denary	0	1	2	3	4	5	6	7
Binary	0000	0001	0010	0011	0100	0101	0110	0111
Hex	0	1	2	3	4	5	6	7

Denary	8	9	10	11	12	13	14	15
Binary	1000	1001	1010	1011	1100	1101	1110	1111
Hex	8	9	A	B	C	D	E	F

For example, octet 01001110 can be considered as two half-octets 0100 and 1110, and from Fig. 11.5 these can be represented in hex code as 4 and E respectively. So the original octet 01001110 can be represented more simply as 4E in hex form. This is of great value in many computer and data communications applications, with all the 256 possible octets from 00000000 to 11111111 being represented by hex in the range 00 to FF.

Standarization of computer codes With the rapid growth of computer applications and data services throughout the world, a number of different codes were developed simultaneously, and it became obvious that a standard was required for the international exchange of data information. In 1962, the United States proposed their American Standard Code for Information Interchange (**ASCII**). This uses a 7-bit data byte with 128 (2^7) possible characters. The 7-bit data byte is normally followed by a parity-check bit for error detection. This parity bit is chosen to give either an odd number of 1s (odd parity) or an even number of 1s (even parity) in any octet. The receiving equipment is designed to detect errors due to any one single bit in each octet. There is, however, no provision in the protocol to tell the sending end terminal that an error has been detected.

The problem of code standardization was considered by the International Standards Organization (ISO), the International Telegraph and Telephone Consultative Committee (CCITT) and various national organizations, and in 1968 the International Alphabet No. 5 (IA5) was formally agreed. It is virtually the same as ASCII, with provisions for minor national variations.

The other main code in use, particularly for IBM computers and terminals, is the Extended Binary Coded Decimal Interchange Code (EBCDIC), which uses 8-bit bytes with 256 (2^8) possible characters.

For data bytes transmitted in serial form, the bit rates at the transmitting and receiving machines or terminals are controlled by internal timing devices called **clocks**. The bit transmission can be *asynchronous* (anisochronous) or *synchronous* (isochronous).

Asynchronous Transmission

Start and stop bits are placed before and after the information byte bits as previously described for teleprinter telegraphy. The start bit from the transmitting terminal switches on the clock in the receiving terminal which controls the sampling of the byte information. The stop signal switches off the clock. So the receiving terminal clock is restarted at the beginning of each byte received, and there can be a gap of any length of time between bytes. Asynchronous transmission is illustrated in Fig. 11.6.

Fig. 11.6 Asynchronous serial data bit transmission

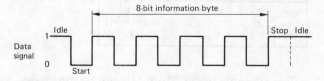

Asynchronous transmission is relatively easy, but it is somewhat inefficient because at least 10 bits have to be transmitted for each 7-bit or 8-bit information byte.

Synchronous Transmission

No start and stop bits are used, and the receiving terminal clock runs continuously. For correct operation, the receiving equipment must start to sample the incoming line data signal on the first bit of each character byte, or it will be out of step with the transmitted signal, and each successive byte will be detected incorrectly. To achieve synchronization between the transmitting and receiving terminal clocks, each stream of data bytes is preceded by two or more synchronization (SYN) bytes which have a predetermined bit pattern recognized by the receiving terminal to identify the first subsequent data bit. The importance of clock synchronization is illustrated in Fig. 11.7.

Fig. 11.7 The importance of clock synchronization

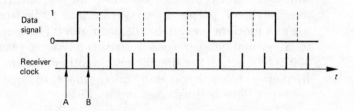

If the receiving terminal clock begins the sampling process at time A, the 8-bit character byte would be interpreted as 01100110, but if the sampling commences at time B the interpretation would be 11001101.

Computer/Terminal Interconnections

For intercommunication between two computers, or between a terminal and a computer, over short distances, simple private lines can be used. This applies to links within a building, or between closely located buildings. A line connection can take one of three possible forms:

1 *Simplex*, where transmission can occur in only one direction.
2 *Half-duplex*, where transmission can occur in either direction, but only in one direction at a time.
3 *Full-duplex*, where transmission can occur in both directions simultaneously.

The binary digits could be represented by a single-polarity voltage level and zero, giving single-current operation. For a number of reasons that give rise to severe distortion of single-current pulses, the binary digits are usually represented by reversing a voltage potential to give double-current operation. For short distances, data bits are typically represented by ±6 V pulses, as illustrated in Fig. 11.8 as a simple on-line connection.

Fig. 11.8 On-line computer/terminal connection

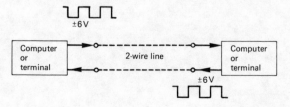

Signalling distance can be increased by increasing the level of the double-polarity voltages, but there is still a severe limitation on the maximum bit-rate that can be achieved due to signal distortion caused by line resistance and capacitance. For external line connections of up to 3 or 4 km, the maximum signalling speed is about 150 bit/s. If necessary, the distance over which the ± 6 V signals can be successfully transmitted can be increased by the use of *regenerators* spaced at regular intervals along the line, as previously explained in Chapter 10 for pcm systems.

Use of Modems

A fairly simple solution to the distance problem can be achieved in a way similar to that previously described for VF telegraphy. The computer/ terminal ± 6 V pulses are converted into audio-frequency tones by the use of a **modem** (modulator/demodulator). The tones can then be passed to a normal analog speech circuit having a working bandwidth of 300–3400 Hz. At the distant end, another modem converts the audio-frequency tones back into ± 6 V pulses to apply to the computer/ terminal. This is illustrated in Fig. 11.9.

Fig. 11.9 Simple use of modems for data communications

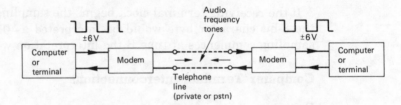

The analog speech line can be a private one leased from the telephone administration as a permanent line, and called a **dedicated circuit**. Alternatively, it is possible to set up a connection as and when desired by dialling over the public switched telephone network (pstn).

Although modems are becoming increasingly sophisticated and capable of high transmitting speeds, the overall signalling speed is limited by the characteristics of the analog telephone line. For faster signalling speeds and more flexible interconnection, a purely digital network is clearly desirable. Rapid developments have been made to this end, and will be considered in the next chapter.

12 Introduction to Digital Networks

Introduction

The very earliest digital network could be considered to be the public telegraph network using manually-operated teleprinters that generated the IA2 telegraph code at 50 bit/s as described in Chapter 11. The original network was manually operated and switched with telegrams being relayed from switchboard to switchboard until the destination was reached. This network was eventually replaced by an automatically-switched network in which the originating telegram operator could establish a dialled connection to the destination switchboard, and the telegram was then transmitted from end-to-end instead of being relayed manually as hitherto.

A public teleprinter network called Telex was also established. Originally, Telex subscribers had a teleprinter installed alongside their telephone instruments, calls were set up to other Telex subscribers over the public telephone network, and then both originating and receiving subscribers switched to teleprinter and the message was transmitted. Later, an automatically-switched public Telex network was introduced quite separate from the existing public telephone network, and eventually this was extended into a world-wide network that is still widely used today, but still at low data rates.

It is worth remembering at this point that both the telephone and telex networks were designed to cater for signals at the information handling rates of people, either by speaking or operating a teleprinter keyboard. Improved teleprinters and lines have now made it possible to introduce a new world-wide "super telex" network called Teletex capable of handling messages at data rates of 2.4 kbit/s.

It was seen in Chapter 11 that interconnection between computer and terminals is possible by directly-connected private leased circuits, as illustrated in Fig. 11.8. Over very short lines data transmission rates of up to 9.6 kbit/s are possible. Longer links are possible by use of modems, either by directly-connected private leased lines or by setting up a dialled connection over the pstn, followed by simultaneous switching to data working at both ends of the link. This is illustrated in Fig. 12.1.

The simple arrangement in Fig. 12.1 is of course very similar to the original Telex network previously described, where the pstn could be

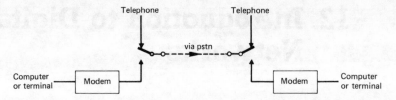

Fig. 12.1 Principle of data link over the pstn

switched at each end between either telephone instruments or teleprinters. Once the connection is established, it is exclusive to the two users concerned, so the network is called **circuit-switched**. This is clearly suitable for relatively infrequent communication between the two users. Also, users can avail themselves of the cheap-rate facility of the pstn if convenient, since the call is charged on time and distance as for a normal telephone call. However, users have to tolerate the normal problems of delays due to engaged destinations and lines, and of course expensive peak-rate charges, etc.

Datel Services in the UK

In the early 1960s the British Post Office (now British Telecom) introduced a number of data communication services under the general name of Datel. The facilities and capacity of these services were graded in different tariffs, with better performance costing progressively more. The development of these services is illustrated in the following summary (for CCITT references, see page 128).

Datel 100 (CCITT V10), 1964

Private telegraph circuits, tariff H 50 bit/s, tariff J 110 bit/s, public Telex network 50 bit/s – all these services are asynchronous.
Datel 200 (CCITT V21), 1967
Public telephone network or private leased circuit, 200 bit/s, asynchronous.
Datel 400, 1973
PSTN or private leased circuit, 600 bit/s, asynchronous, one-way only, outstation to central.
Datel 600 (CCITT V23), 1965
PSTN or 4-wire private leased circuit, up to 1200 bit/s asynchronous.
Datel 2400 (CCITT V26A), 1968
Private leased circuit, 2.4 kbit/s, synchronous.
PSTN (standby), 1.2 kbit/s, synchronous.
Datel 2400 dial-up, 1972
PSTN, 600 bit/s assured (2.4 kbit/s possible), synchronous.
Datel 48K, 1970
Special-quality leased circuit, 40.8 to 50 kbit/s, synchronous.

CCITT Recommendations	Data rate bit/s	Corresponding highest modulating frequency f_{max}	Transmitted frequencies Hz 0	Transmitted frequencies Hz 1	Nominal carrier frequency Hz	Frequency deviation of nominal carrier
V21	up to 300	150 Hz	1180 1850	980 1650	1080 1250	±100 Hz
V23	600	300 Hz	1700	1300	1500	±200 Hz
V23	1200	600 Hz	2100	1300	1700	±400 Hz

Digital Modulation

The function of a modem in a digital network is to change the data signals (e.g. ±6 V) from computers and other data terminal devices into audio-frequency tones for transmission over telephone lines. This can be achieved in a number of ways.

Frequency-shift keying or modulation (fsk) An audio-frequency carrier is used having one value for a 0-bit and another value for a 1-bit, for each direction of transmission. The ±6 V data signals control the frequency of an *LC* oscillator or multivibrator between predetermined limits, as illustrated in Fig. 12.2.

Fig. 12.2 Digital FSK waveforms

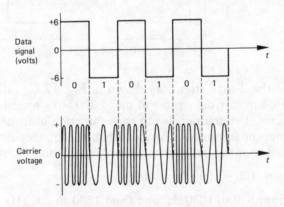

The bandwidth of a normal telephone line is from 300–3400 Hz, but not all of this is available for audio-frequency data communication because of the inband signalling and supervisory facilities needed on the pstn. The bandwidth available is therefore restricted to the bands 300–500 Hz and 900–2100 Hz.

For a frequency-modulated waveform without distortion, the bandwidth needed to accommodate the waveform is given by the empirical formula:

$$\text{Bandwidth} = 2 \times \left[\begin{array}{l} \text{Frequency deviation of the nominal carrier } f_d \\ \textit{plus} \text{ Highest modulating signal frequency } f_{max} \end{array} \right]$$

It was shown in Fig. 6.9 that for telephone circuits longer than 32–40 km (20–25 miles), a 4-wire arrangement is used to enable amplification to be applied in both directions of transmission to overcome the effect of line attenuation. For data communications, if the modems replace the telephone instruments in the 2-wire ends, a 2-wire data circuit results. But if the hybrid terminating units are replaced by modems, a 4-wire data circuit is achieved. This is illustrated in Fig. 12.3.

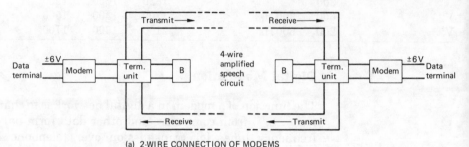

(a) 2-WIRE CONNECTION OF MODEMS

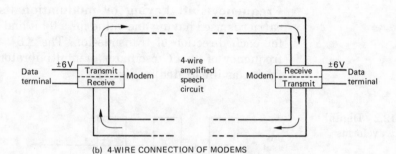

(b) 4-WIRE CONNECTION OF MODEMS

For the 2-wire circuit as shown in Fig. 12.3a, full duplex operation is possible at low data rates of up to 300 bit/s by using four different frequencies (two frequencies in each direction of transmission).

From the formula for bandwidth of a frequency-modulated wave, it can be seen that the bandwidth needed for this low data-rate transmission is

$$2 \times (100 + 150)\,\mathrm{Hz} = 500\,\mathrm{Hz}$$

i.e. from 830 to 1330 Hz, and from 1500 to 2000 Hz.

For the 4-wire circuit as shown in Fig. 12.3b, full duplex operation is possible at data rates of 600 and 1200 bit/s by using only two frequencies.

For 600 bit/s data rate, the bandwidth needed is given by

$$2(200 + 300)\,\mathrm{Hz} = 1000\,\mathrm{Hz}$$

i.e. from 1000 to 2000 Hz.

For 1200 bit/s data rate, the bandwidth needed is given by

$$2(400 + 600)\,\mathrm{Hz} = 2000\,\mathrm{Hz}$$

i.e. from 700 to 2700 Hz. This would seem to exceed the available bandwidth of 900 to 2100 Hz as quoted previously. In fact, all the harmonics present in the ±6 V square-wave data signal are not absolutely necessary, and in practice a bandwidth of 1200 Hz (equal to the bit rate) is found to be adequate.

Phase modulation The simplest form of digital phase modulation is called **phase inversion** (i.e. 180° phase shift) where the phase of a carrier frequency for a 1-bit is 180° out of phase with that for a 0-bit, as illustrated in Fig. 12.4. This method is often called *two-phase modulation*, and detection by a receiver requires the recognition of the two different phases to reproduce the original digital waveform.

Fig. 12.4 Simple phase inversion (two-phase) digital modulation

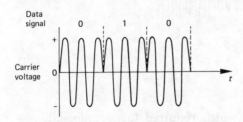

Another method called **differential phase modulation** uses a phase change of the carrier relative to that for the previous bit. At the same time, in order to reduce the effective bit rate and also therefore the bandwidth required, successive bits in a data stream are considered in *pairs* or *dibits*. There are only four possible combinations of dibits, that is 00, 01, 10 and 11. Each of these combinations can be represented by a particular value of phase-shift relative to the previous bit, thus giving a possible 4-phase signal. There are two alternative phase-shift values recommended by CCITT V26, as follows:

Dibit	00	01	10	11	Carrier frequency
Phase change A	0°	+ 90°	+180°	+270°	1800 Hz
or					
Phase change B	+45°	+135°	+225°	+315°	1800 Hz

With alternative A, the absence of any phase change for a long stream of consecutive zeros can cause problems with synchronization, but this does not arise with alternative B.

Amplitude modulation The simplest form of digital amplitude modulation is to arrange for each 1-bit to switch the carrier off and for each 0-bit to switch the carrier on at a fixed voltage level, as illustrated in Fig. 12.5. Alternatively, each 1-bit can be arranged to give a fixed *lower* amplitude level of carrier voltage than for each 0-bit, instead of zero voltage, as shown in Fig. 12.5.

Fig. 12.5 Simple digital amplitude modulation

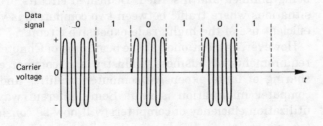

Fig. 12.6 Multi-level digital amplitude modulation

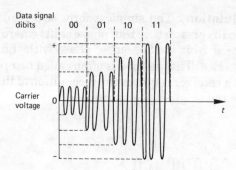

The bandwidth required to accommodate the digital-modulated am signal can be reduced by representing the original data stream as successive dibits, as previously explained for phase modulation, and then arranging for each of the four possible dibit combinations to produce four different levels of carrier voltage. This is illustrated in Fig. 12.6.

Vestigial sideband modulation (vsb) In Chapter 4 the principle of vestigial sideband amplitude modulation was introduced as a means of reducing the bandwidth required for a television signal, but at the same time retaining the relative ease and low cost of demodulating a double sideband amplitude-modulated signal.

This principle is also used for high-speed data transmission at 48 kbit/s using a high-quality leased line and BT modems to connect two computers or data terminal equipments. 48 kbit/s represents a highest fundamental frequency of 24 kHz, but some lower harmonics are also transmitted, giving a maximum transmitted frequency of about 36 kHz. By normal DSBAM transmission a bandwidth of 72 kHz would therefore be necessary, but by using VSBAM techniques the bandwidth required can be reduced to about 44 kHz.

As already illustrated in Fig. 10.5, twelve conventional speech channels can be transmitted simultaneously on a high-grade bearer in a frequency band of 60–108 kHz. Clearly this grade of bearer would therefore also be capable of transmitting the VSB digital signal at 48 kbit/s mentioned above, using a carrier frequency of 100 kHz and a total digital modulated bandwidth of about 40 kHz.

Private modems are now available which enable data to be transmitted at rates of up to 168 kbit/s over high-grade leased (dedicated) circuits, but these are likely to be superseded by the new digital services and networks being planned and installed. Dedicated circuits are obviously ideal for situations where traffic between two computers is very frequent, giving efficient use of the high-grade expensive circuit.

However, as mentioned at the beginning of Chapter 11, the present-day requirements of business, industry and commerce are such that the sharing of large expensive computer facilities and the interchange of computer information is vital. Some different ways of improving the utilization efficiency of computers will now be considered.

Simple Networks

Point-to-point network   This is a simple example of leased lines connecting more than one terminal equipment to a central computer, and is illustrated in Fig. 12.7. Each terminal has immediate access at all times,

Fig. 12.7   Point-to-point data network

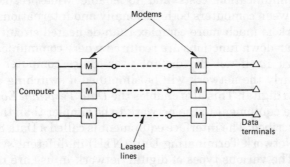

with minimum transmission delays. The computer software is simple and relatively cheap. Each terminal has exclusive access to the data-rate capacity of its own leased line, but there could be long idle periods so the method tends to be inefficient.

Polled network   A number of terminals use the same leased line as illustrated in Fig. 12.8. This method gives more economic use of the line

Fig. 12.8   Polled data network

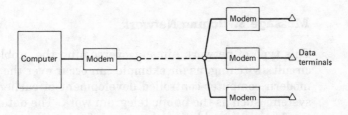

capacity but the computer software is more complex and costly. The computer has to interrogate each terminal in turn to see if it wishes to transmit data. This is called **polling**, and must introduce delays if more than one terminal wishes to transmit simultaneously.

Multiplexing   Using fdm or tdm techniques, as described in Chapter 10, several low-rate data signals can share one high-rate data circuit. For example, four 300 bit/s data signals can be carried simultaneously on one high-cost line capable of handling data at 1200 bit/s. This is illustrated in Fig. 12.9.

Fig. 12.9   Sharing of computer by multiplexed terminals

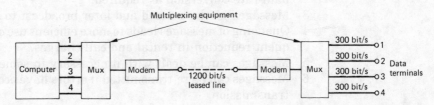

Digital Networks

Computers are capable of handling information at data rates very much higher than is possible using modems and audio-frequency signals over telephone lines. A digital network is clearly needed to minimize data communication costs and to enable widespread intercommunication between computers both nationally and internationally. Obviously a network is much more complex than dedicated circuits because set-up and clear-down functions are required where communication is with the network itself not the distant user. Since computer data is digital then clearly the network will be simplified if switching and transmission are also digital. This also enables the *Data Terminal Equipment* (DTE) interface equipment to the network to be simpler than the modems previously described. The interface equipment is called a Data Service Unit (DSU) or a Network Terminating Unit (NTU) in different systems.

The various types of digital network in use are now described.

Circuit Switching Network

As in an example described previously, the pstn provides circuit switching of data signals. Two terminals (DTE) are connected exclusively as required by a real-time circuit. The destination terminal may be engaged, lines in the pstn may be engaged, and the call may eventually be established during peak call charges.

Message Switching Network

This type of network aims at overcoming the problems encountered by circuit switching, as for example can occur over the pstn, and is really a modern computer-controlled development of earlier message switching systems such as the public telegram work. The data message containing the destination address or location is fed into the system where it is stored and subsequently forwarded when the destination is free, or at an appropriate charging time. For this reason, the network is sometimes called a **store-and-forward** network. Clearly, delays in transmission must be tolerated and interactive communication between terminals is not possible. The following features are typical of message switched networks:

1 Messages can be entered whether the destination terminal is free or engaged.
2 Terminals using different codes and different data rates can communicate, since the system can be designed to perform automatic code and data-rate conversion as required.
3 Messages can be entered and later broadcast to several terminals.
4 Queueing of messages leads to more efficient use of circuits with subsequent reduction in rental and call charges.
5 Messages can be dealt with if necessary in some order of priority.
6 Messages can be re-transmitted if errors have occurred in the original transmission.

Packet Switching Network

This type of network was introduced in the USA in 1964, and in other countries in the early 1970s. Each network had different characteristics with no possible interchanging of information. A dedicated link is not switched between two users for the duration of a complete message. Instead, data messages are divided into discrete quantities called **packets**, each of which has a maximum length and specified format. Since there are no long data streams, blocking or congestion of links is less likely, so delays are usually small. Consequently, interactive working between users is possible.

A packet typically contains up to 128 octets of one user's data signal, and is sometimes called a *datagram*. It contains identification of the destination terminal, information concerning the position of the packet in the complete message, and other control data which makes error-detection possible.

The data packets are collected together and separated by small computers which are individual to each sender and receiver, and are called **packet assemblers/disassemblers** (PAD).

This type of network is called a Packet Switched Service (PSS) or a Packet Switched Digital Network (PSDN). In the UK, British Telecom have been concerned with packet switching since 1973, beginning with an Experimental Packet Switched Service (EPSS), and later with Euronet and the International Packet Switched Service (IPSS). Then in 1981, BT opened a national public network called the Packed-Switched Service (PSS). This network has a national management centre (NMC) in London where duplicated minicomputers are installed, each one having 600 Mbytes of disc storage. The network has access to the International Packet Switched Service (IPSS) and also includes links to Slough, Reading, Bristol, Cambridge, Birmingham, Liverpool, Manchester, Leeds, Newcastle, Glasgow and Edinburgh.

Local Area Network (LAN)

LANs are networks permitting the interconnection and intercommunication of a group of computers, primarily for the sharing of resources such as data storage devices and printers. Business information (e.g. speech, data, text and graphics) can be directed wherever desired within a building, or within a local area of up to 3 or 4 kilometres. Different LANs can intercommunicate with each other, or be connected to public networks by using **gateway** interfaces. Local wiring is installed privately using twisted pairs, multicore cables, coaxial cables or optic fibre cables. Some LANs have even used the wiring already existing for PBX telephone extensions.

Since high data rates are needed for computer data interchange, or for several computers to share a very expensive common resource such as a large disk store or very-high-speed printer or plotter, LANs can be designed to operate at speeds of up to 10 Megabit/s (Mbit/s). Speech transmission can also be handled at 64 kbit/s per telephone conversation.

However, due to the complexity and high cost involved and with a distinct lack of international standards and user familiarity, development of high data-rate LANs has been slow; by comparison, the introduction of low-cost low-rate LANs is progressing more rapidly.

A brief summary of the development of LANs will now be considered. Most LANs can be classified as Star, Ring or Bus networks.

Star network A central control unit contains storage and switching equipment to enable a number of other stations to intercommunicate completely as and when desired. This is illustrated in Fig. 12.10, with data or speech transfer indicated from X to Y.

Fig. 12.10 Star data network

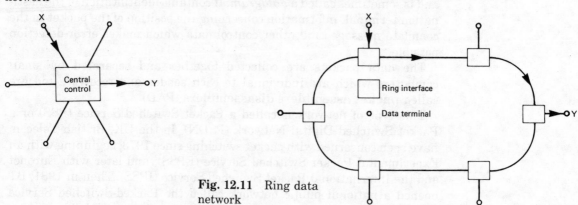

Fig. 12.11 Ring data network

Ring network Storage and switching control equipment is distributed between various data devices which are all connected to a continuous ring connection, as illustrated in Fig. 12.11, with data transfer indicated from X to Y via an intermediate station. Each terminal has a ring interface equipment that applies the data signal to the ring at high data rates with low error rate. Each interface includes an amplifier to boost the signals applied to the ring.

Fig. 12.12 Bus data network

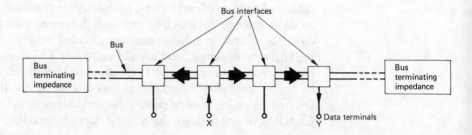

Bus network Again, storage and switching control equipment is distributed between the various devices connected to a common highway or bus connection, with all devices having equal status. A typical bus network is illustrated in Fig. 12.12.

With ring and bus networks, there is no polling of terminals, and any terminal can send data at any time once it has established that the ring or bus is free. If two terminals start simultaneously, a **collision detection** occurs and a jamming signal is sent to stop all devices. Each device then listens again for a free condition. The data received on the collision detection is ignored, so there is no corruption of the signal. This form of access is called Carrier Sense Multiple Access with Collision Detection (CSMA/CD).

Other techniques that have evolved are *empty slot* and *token-passing* (token access).

With **empty-slot** transmission, mini-packets of data circulate around the ring or along the bus. These mini-packets contain a start bit, a full/empty bit, destination address, source address, data bytes, control bits, response bits and a parity bit. When a terminal wishes to transmit data it must wait for an empty mini-packet to arrive. It then completes the data concerned and passes it on to the next terminal, and so on, until the destination terminal identifies and seizes the packet.

With **token-passing**, a token circulating around the ring or along a bus indicates a free network. A terminal wishing to transmit removes the token, transmits its data, and then replaces the free token to the network. The token for any particular network is a predetermined bit pattern that is not allowed to occur in any data stream.

Examples of LANs

Private Automatic Branch Exchange (PABX)   This can be regarded as a typical example of a star-type LAN. The use of manual and automatic private branch exchanges by large organizations in the national telephone network is no doubt very familiar to all. They allow a relatively small number of public exchange lines to be shared by a large number of internal extensions and also allow internal extension intercommunication.

The present digital transmission and switching techniques have enabled a new generation of PABXs to be developed that can be totally digital for internal speech and data transmission (but not text and graphics). They can also intercommunicate with the existing analog public speech network, or with private digital data networks.

Ethernet   This is a bus-type 10 Mbit/s LAN developed initially in the early 1970s by Xerox, and later joined by the computer company Digital Equipment Corporation (DEC) and chip manufacturers Intel in an attempt to achieve a standardized CSMA/CD technique. Ethernet is based on 500-metre lengths or segments of coaxial cable bus, called the Ether, that are correctly terminated at each end. The cable can be tapped at any 2.5 metre point along its length to connect up to 100 transmitter/receivers. Each transceiver can then be connected to a terminal by a cable of up to 50 metres in length. This arrangement allows flexibility of terminal location without necessarily having to re-route the coaxial bus. Several of these 500 metre lengths or segments can be interconnected by use of repeaters, as illustrated in Fig. 12.13.

Fig. 12.13 Principle
of Ethernet

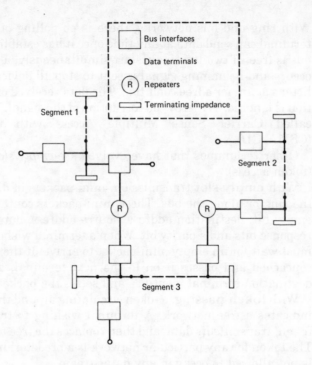

Bus interfaces

o Data terminals

(R) Repeaters

Terminating impedance

Segment 1

Segment 2

R   R

Segment 3

The Cambridge Ring This is a ring-type 10 Mbit/s LAN which was developed by Cambridge University from 1974 onwards for interconnecting a large range of computers and other terminal devices. The network uses the circulating empty-slot technique previously described. The ring can consist of an ordinary telephone pair, coaxial cable or optic fibre cable, and each type requires a particular repeater or ring interface spacing (for example, 100 to 300 metres for telephone cable; 2 km for optic fibre cable). Further development is aimed at achieving 100 Mbit/s data transmission using specially manufactured Cambridge Ring integrated chips.

All-Digital Services

In the UK, three organizations have been granted Government licences to go ahead with the provision of digital networks. They are British Telecom (BT), Mercury Communications (Cable and Wireless) and the City of Hull (historically, independent of the national telephone system).

BT have introduced a series of services under the heading of *X-Stream Services* which are now briefly summarized.

Switch Stream One (Packet Switched Service, PSS)
This *packet-switched* public data service was introduced in August 1981, having a very fast set-up that is ideal for computers, and terminals located at typical sales points. International facilities were added in early 1982.

Switch Stream Two
Introduced in 1983 as a *circuit-switched* digital network suitable for processing of remote ordering, credit verification and typical on-line data retrieval.

Megastream
Introduced in 1982/83 as a point-to-point service which can be connected directly to modern digital PABXs for speech transmission, or it can be used for very-high-speed data transmission at 2 Mbit/s or 8 Mbit/s. The 2 Mbit/s service can carry 30 telephone conversations simultaneously, or data, or a mixture of speech and data. An international service was added in 1983/84.

Kilostream
This service was introduced in 1982, became available nationally in January 1983, and internationally in 1983/84. It offers digital services at 2.4, 4.8, 9.6, 48 and 64 kbit/s, as follows:

2.4 kbit/s – data links between VDUs, teleprinters or teletypewriters for electronic mail, etc.

4.8 kbit/s – credit verifications, slow-scan tv.

9.6 kbit/s – high-speed facsimile.

48 kbit/s – high-speed computer data transmission.

64 kbit/s – speech transmission.

Sat Stream
Introduced in 1984 to offer private satellite links using small dish aerials that beam signals to the European Communications Satellite and Telecom 1. The very flexible service will provide a range of digital transmission speeds that will be useful for remote locations such as off-shore oil platforms, etc. Copies of newspapers can be sent many times faster than hitherto, so that companies with multi-point facilities will be able to publish editions simultaneously in cities throughout Europe. News Agencies will be able to distribute information to many different locations simultaneously, as will the large banks, oil companies and other multinational organizations.

Mercury Communications
Mercury Communications are planning and installing an all-digital national network using a combination of microwave radio links and optic fibre cables.

Project Universe
This is a project that includes BT and others which is investigating the possibility of providing intercommunication between LANs on a national and international basis. The project is called the UNIVersities Extended Ring and Satellite Experiments (UNIVERSE), and began in 1981. It uses an Orbital Test Satellite (OTS) launched in 1979 by the European Space Agency (ESA) to investigate satellite communications in the 11–14 GHz band.

Value-Added Network (VAN)

This is a scheme whereby a third party can offer to users, by means of private terminal equipment, a range of point-to-point or switched services over and above those offered by the administrators of public digital networks such as BT or Mercury. The third party, by Government licencing regulations, must add some extra value to the normal facilities offered by the bearer. The type of service envisaged or already offered includes data protocol conversion, store-and-forward message facility, links to other networks, home banking, holiday reservations, retail invoicing, stock control, library services, weather forecasts, facsimile, and electronic mail. *Electronic mail* is a service where messages can be sent to any user having telephone or computer at any time. If the destination user is not in, messages can be stored in an 'electronic mailbox' for transmission later on. A wide range of private terminal equipments can be used, e.g. personal computers, word processors, and Prestel equipment.

There are three categories of network either already in operation or being developed. The first category offers data storage and processing to a wide range of users. An example of this is the Comet electronic mail service which operates over either British Telecom's packet-switched network or any private network. Messages are sorted and stored in a central computer, and access can be obtained by terminals that are either directly connected by leased line or by a telephone fitted with an acoustic coupler and modem.

The second category offers specific service to a particular group of users, such as banks and building societies. An example of this is the Homelink system operated by the Nottingham Building Society which links customers to the society, the Bank of Scotland, and a selected range of shopping services. A special console supplied by the society operates with a normal television receiver. The system is available for 18 hours a day for all seven days of the week, and it enables customers to

View a statement of their account.
Transfer money from a bank to building society, and vice versa.
Apply for a mortgage.
Obtain a quotation for a loan.
Make holiday reservations.
Shop from home, usually at stores offering discount prices.

The third category offers facilities for publication of information by large organizations.

There are at present a large number of VANs in operation using different systems of architecture and protocol. This makes intercommunication between systems virtually impossible without some suitable conversion interface.

BT and IBM proposed a joint VAN system recently, but this was blocked by the Department of Trade and Industry. Despite this setback, IBM are still planning to proceed alone with their project which uses their own Systems Network Architecture (SNA). This will of course require suitable conversion interfaces to communicate with other systems.

Other companies, however, recognized the need to have a standard architecture and protocol to simplify complete international data communication. The problem was taken up by CCITT and the International Standards Organization (ISO) to produce a standard architecture and protocol called **Open Systems Interconnection** (OSI). This will contain a set of standards ranging from the simplest factors, such as the number of pins on a plug, to the way in which complex programs will work together. There will be seven different sub-sections or *layers* dealing with the various aspects of computer communication which make up the complete system. These are the physical layer for connection and disconnection of the link, the datalink layer for synchronization and error control, the network layer controlling the switching facilities, the transport layer for user/network interface, the session layer controlling communication between cooperating functions, the presentation layer to interpret the information being exchanged, and the application layer to serve the destination user. This arrangement shows a number of differences compared with the SNA of the IBM system.

Integrated Services Digital Network (ISDN)

This is perhaps the natural outcome of all the recent advances in the use of digital networks. The object is to produce an integrated digital network capable of controlling, switching and transmission by common equipment all possible forms of communications services. In the UK these services include telephone, telex, digital PABXs, Prestel and Picture-Prestel, facsimile, slow-scan tv, computer data services, electronic mail, telemetry of household meter readings for electricity, gas and water, electronic shopping, all of which can provide similar digital data streams.

The analog pstn, with its slow setting-up of connections, low bandwidth and simple signalling, has been a major stumbling block in setting up an integrated services network.

As far as telephony is concerned, an integrated network would involve the integration of digital transmission techniques already in use (e.g. pcm) with the digital switching techniques now being developed and introduced in digital PABXs and System X.

The ISDN can be considered as having two sections. One section is an *Integrated Digital Network* (IDN) of System X exchanges interconnected by high-capacity data links. The other section is an *Integrated Digital Access* (IDA) of single or multi-line local systems to connect all types of terminal equipment to the IDN. The IDN allows very fast set-up of end-to-end connections, and the IDA provides all users with a digital interface without the need for expensive modems.

The ISDN will initially be linked by suitable interfaces to existing services such as pstn, packet-switching and X-stream services, etc. to allow users to continue operating their existing equipment for a reasonable economic period. Connections will of course be possible to digital networks in other countries.

The principle of an ISDN network is illustrated in Fig. 12.14.

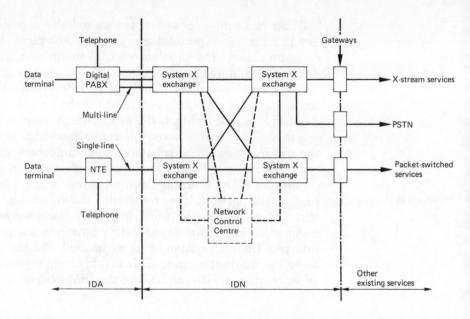

Fig. 12.14 Simple illustration of ISDN

Standardization

In order to try and standardize technical details for data communications throughout the world, the CCITT (Comité Consultatif International Télégraphe et Téléphone) have produced a number of recommendations, after international discussions.

The **V-series** of recommendations covers data transmissions over normal telephone circuits, and a few of these recommendations are now listed:

    V3 – IA5 code.
    V4 – structure of IA5 signals.
    V21 – 200 baud modem for use on the pstn.
    V24 – list of definitions for interchange circuits between data terminal equipment and data circuit terminal equipment (modem).
    V26 – 2.4 kbit/s modem for use on 4-wire point-to-point circuits.
    V29 – 9.6 kbit/s modem for use on leased circuits.
    V57 – comprehensive data test-set for high data signalling rates.

The **X-series** of recommendations covers data networks, and a few of these recommendations are now listed:

    X3 – PAD in a public data network.
    X25 – Interface between DTE and data circuit terminating equipment for terminals operating in the packet mode on public data networks.
    X28 – DTE/DCE interface for asynchronous (stop/start) mode DTE accessing the PAD in a public data network in the same country.

13 Public Information Services – Teletext and Prestel (Viewdata)

TELETEXT: Introduction

On page 34, Chapter 4, it was explained that in any tv system not all the lines of the scan are used to carry picture information. Some of the line periods are used to send field synchronizing pulses, and others are kept *blank* to ensure that the scanning electron beam returns to the correct position to begin the next field. For a long time, tv engineers have considered whether the blanking periods could be used to transmit other information signals.

The first attempt was to transmit test signals to enable the performance of a tv system to be checked during programme transmissions. Next came the introduction of data signals (see Chapter 11) by the IBA to identify and label programmes from the various companies in the Independent TV network. This was known as SLICE (Source Labelling Indication and Control Equipment). A similar system was used by the BBC, and this is still in use in the Eurovision network.

The next logical step was to consider using the line blanking periods to transmit a variety of information in data form, so that tv viewers could use such information as required by adding a decoder unit and a selector unit to their normal tv receiver. So, in the early 1970s, both IBA and BBC began the development and testing of similar data information services.

The IBA system was called ORACLE (Optional Reception of Announcements by Coded Line Electronics). This consisted originally of about 50 pages of information, each page having 22 lines of text with 40 characters in each line. At first only black and white reproduction was possible, but later colour facility was added.

The BBC system was originally called TELEDATA, but was later renamed CEEFAX (see facts). This consisted of about 32 pages of information, each page having 24 lines of text with 32 characters per line. The data rate (see Chapter 11) was faster for Ceefax than for Oracle.

It soon became obvious that, in the interests of all concerned, some *standard* system having certain technical specifications was necessary. The IBA, BBC and the TV industry (BREMA) then worked together to produce the specifications of a standard system, which was called TELE-TEXT, published in October 1974. However, the IBA and BBC retained their original individual titles of Oracle and Ceefax respectively within this common specification.

Teletext Specification

Lines 17, 18, 330 and 331 are used to carry the data signals.

Up to 800 pages of information are available, arranged in 8 magazines of 100 pages each.

Various colours are available as well as black and white, and graphic designs are also available, e.g. weather maps, etc.

Each page consists of a page *header* and 23 lines of information of 40 characters each. The page header gives the name of the service (Oracle or Ceefax), the page number, the date and the time.

Since 4 lines of information are transmitted per complete tv picture (every 2 fields), then 12 fields are needed for each page of teletext information. Now fields are transmitted at 50 per second, so each field takes 20 milliseconds.

The 12 fields needed for each page of teletext therefore take 240 milliseconds. So the teletext information is transmitted at approx. 4 pages per second.

The data signals are transmitted at the rate of 6.9375 Mbits/sec.

Receiving Teletext

To receive the teletext information signals, a normal domestic tv receiver requires an *additional decoder unit*, and a means of *selecting* the information page required from all the available pages being transmitted along with the normal tv signal.

The simple arrangement of a teletext system is illustrated in Fig. 13.1, and a simple block diagram of a modified tv receiver is shown in Fig. 13.2.

Since the viewer can only select from the complete range of information being transmitted with the normal tv signal, teletext is essentially a *one-way* communication system. Also, it takes a little time for the selected page to be recognized and displayed (up to 24 seconds).

PRESTEL: Introduction

While the Teletext system was being developed for the tv broadcast networks, the British Post Office (now British Telecom) was developing an alternative information service for the national telephone network, called VIEWDATA. Information is stored in various computer stores, and is available by access from a normal telephone user if a data modem is provided together with a modified tv receiver or other commercial terminal display unit. The system was later renamed PRESTEL.

Simple Principles of Prestel

The available information is stored in local, regional and national computer centres, as illustrated in Fig. 13.3.

The Prestel user gains access to the local computer centre by dialling the appropriate telephone number, and when connection is established the user demands the required page of information. Prestel is therefore a

Fig. 13.1 Simple illustration of Teletext system

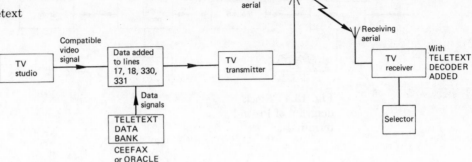

Fig. 13.2 Simple block diagram of modified TV receiver for Teletext

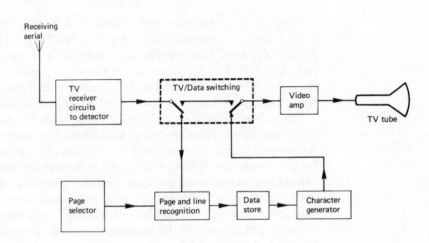

Fig. 13.3 Block diagram of Prestel network

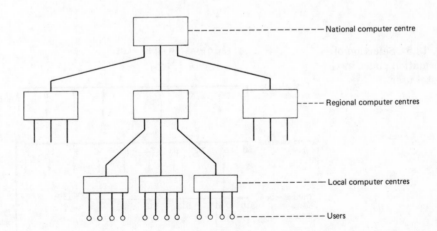

two-way system (*or interactive system*) as opposed to the *one-way system* provided by Teletext.

Further, since only the *selected* information is transmitted from the computer stores on demand, instead of all the available information as in Teletext, the data rate for Prestel can be *slower* than for Teletext. The data signals can therefore be carried by ordinary telephone lines.

Fig. 13.4 shows a simple block diagram of a Prestel terminal.

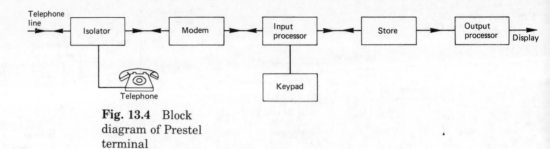

Fig. 13.4 Block
diagram of Prestel
terminal

When a local computer centre is dialled from a Prestel terminal (by a fully automatic process) the computer sends back a 1300 Hz signal to the Prestel decoder. The decoder then returns a signal to the computer to indicate that it is ready to accept the data signals. The computer then requests identification of the terminal for charging purposes, and the terminal replies by sending the particular code of identification.

The computer then sends Page 0, which is the Master Index Page listing the main topics of information available. The user selects one of these main topics, and the computer sends back the appropriate page, which breaks down the chosen main topic into various sub-sections. The user selects one of these sub-sections, and the computer again sends back the appropriate page which further divides the sub-section of the chosen topic into specific detail. This process of one-from-ten selection can be repeated a number of times, and the basic structure of the page selection process is illustrated in Fig. 13.5.

Fig. 13.5 Selection of
information pages by
Prestel user

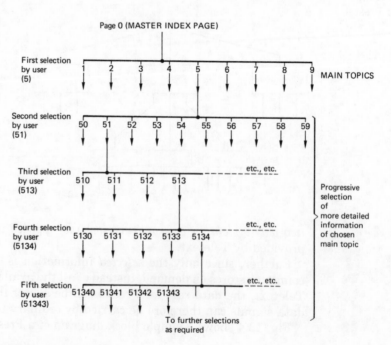

If the *local* computer centre store does not contain the information requested, a connection is automatically made to the *regional* computer centre, or even to the *national* computer centre if needed.

So, a very large amount of information can be stored, e.g. 50 000 pages at a local centre and 500 000 pages at a regional centre, and the user does not need to keep a large index of available information.

Prestel Specification

The data signals use a 10-bit code, made of a start bit, 7 data bits, a parity bit and a stop bit.

The local computer transmits data signals to the terminal modem at 1200 bits/sec, i.e. 120 characters per second, and each page takes approx. 8 seconds.

The terminal modem transmits data signals to the local computer at 75 bits/sec, i.e. 7.5 characters per second.

For 1200 bits/sec signalling, a mark is transmitted as a 1300 Hz tone signal and a space as a 2100 Hz tone.

For 75 bits/sec, a mark is transmitted at a 390 Hz tone and a space as a 450 Hz tone.

When a terminal/computer connection is idling, tones of 1300 Hz and 390 Hz are transmitted.

Prestel can therefore be seen to form a vast library of information with access from any normal telephone instrument if additional equipment is added. It may well be that in the future Prestel users will no longer feel the need for daily or weekly newspapers to keep abreast of news and other wide-ranging information such as sport, entertainment, weather, politics, health, etc.

This may well have significant social effects in due course—only time will tell.

14 Introduction to Cable Television, DBS, and HDTV

Introduction

Television services were originally provided by broadcasting from a small number of high-power transmitters placed at strategic points throughout the country using omnidirectional aerials, with individual domestic receiving aerials and television sets being provided by prospective viewers. These domestic receiving aerials were usually directive, aimed at the transmitting station, except perhaps for locations close to the transmitter. Generally, each transmitting station broadcasts all the different available channels. This arrangement has proved to be satisfactory for many viewers, but the many problems associated with vhf/uhf space-wave propagation has resulted in poor reception for others.

Weak signals arrive at a tv aerial if it is screened from the transmitter by high buildings or hills. It is also possible for several signals to arrive at an aerial due to reflections from high buildings or hills, etc. The distance travelled by each signal can vary, resulting in signals arriving later than others, resulting in "ghost" pictures appearing on the screen. This gives unsatisfactory viewing, even if the directly-received signal is of adequate strength. Further, in remote areas the transmitter may be too far away to give a satisfactory signal level.

One solution provided by the broadcasting companies was to install low-power **booster** (or relay) **transmitters** in problem areas, where the various programmes are re-broadcast on different channels and different aerial polarization to the main high-power transmitter serving the area.

For example, the Midlands is generally served by the 625-line uhf transmitting station at Sutton Coldfield. The region also has over 20 relay stations to serve problem areas. Two of these relay stations and their channels are indicated in the following table:

	BBC 1	BBC 2	ITV	Chan. 4	Power (kW)	Aerial polarization
Sutton Coldfield	46	40	43	50	1000	horizontal
Leamington Spa	56	62	66	58	0.2	vertical
Brierley Hill	57	63	60	53	10	vertical

Community Antenna Television

CATV was another solution to the problems of television broadcasting, provided by local enterprise. It was sometimes called relay tv, but must not be confused with the booster or relay transmitters provided by the broadcasting companies. A communal aerial is installed at some high point, such as a hill or high building where the tv broadcast signals are of adequate strength.

Fig. 14.1 Simplified tree-and-branch cable tv system

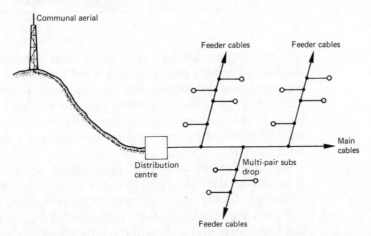

The programmes are then supplied to paying subscribers by cables from a distribution centre. Initially, when only 2 or 3 channels were available, each channel was distributed by a pair of wires, and the subscriber selected a preferred channel by a switch provided on or near the television set. The arrangement is illustrated in Fig. 14.1, which is an example of a **tree-and-branch network**.

Development of Cable Television

The problems of television broadcasting are so acute in the large cities of many countries that the number of subscribers to CATV systems rapidly increased. This was particularly so in the USA with the presence of many skyscraper buildings. More broadcast programme channels became available and at the same time there was an increasing need for sound and television channels carrying specialist and minority-taste programmes, e.g. sport, local interest, classical music, educational, etc. It was apparent that a more sophisticated system was required, such as the use of wideband coaxial cables to replace the "pair of wires per channel" arrangement.

This led to the development of a **star network** as an alternative to the original tree-and-branch network. This is illustrated in Fig. 14.2. The main television centre may receive the available local broadcast channels on a main aerial, or it may initiate these channels directly as well as all the other specialist radio and television programmes. All these programmes are connected by large wideband coaxial cables to various distribution centres, from which smaller coaxial cables with smaller bandwidth are connected to subscribers to the system.

Fig. 14.2 Simplified star cable tv system

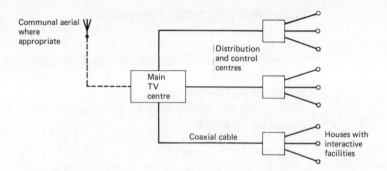

Technological developments now enable the system to be *interactive*, that is subscribers can select a limited number of available channels at any one time by keying information back to its distribution centre. Further, for the very specialist programmes with limited appeal, it is possible to introduce a scrambling facility whereby certain channels can only be received by "coin-in-slot" operation of a descrambler associated with the television set.

The availability of optical fibre cables, with larger bandwidth and lower attenuation than pair or coaxial cables, will enable cable television systems to be extended to much larger areas than hitherto.

Expansion of cable television has varied from country to country, and has been particularly slow in the UK, where it is predicted that only half of the viewing population will have cable tv by the year 2000. There is yet another development which might prove more popular, or may be associated with cable tv services.

Direct Broadcasting by Satellite

Satellite communication was introduced briefly in Chapter 3. The satellite dish aerials transmit a fairly broad beam down to large areas of the earth's surface, so the signal is widely dispersed to give a weak signal at any one point, requiring a large receiving dish aerial at the earth station and an expensive low-noise receiver operating at low temperatures. For example, a typical earth station dish has a diameter of at least 11 metres. Technology has now enabled the satellite transmitter power to be increased, and the dish aerials can produce narrow-beam transmissions. Thus, larger signal levels are now available over smaller geographical areas of the earth's surface, making it possible to broadcast television signals direct to individual houses in a particular area using small dish aerials of about 30–90 cm diameter.

It is planned to use circular equatorial geostationary orbits, with satellites placed at particular points along the orbit to serve different countries. Some satellites could be used to provide broadcasts to one particular country only, whilst other satellites could be shared by several countries to allow an interchange of each other's programmes.

For example, a European satellite, ECS 1, with 5 channels allocated to each country, is already in operation at a longitude of 19°W. At the moment most viewers receive pictures over a cable network, and one channel has been allocated to Britain. A British satellite, Unisat, is being developed, and is scheduled to become operational at a longitude of 31° in 1986. Five channels have been allocated for this satellite, out of the total of 130 channels available to Europe.

One problem at the moment for interchange of television programmes between countries is the fact that there is no international technical standard. The USA uses their NTSC system, France uses their SECAM system, Germany and Britain use the PAL system. At present, planning of DBS systems is based on each country using their own standards, but there seems to be many advantages in trying to introduce one international standard. This seems particularly relevant if one realizes that DBS could become an important source of programmes for cable systems, and a new international system could use new technology to give improved pictures.

High-definition Television

Colour television services at the moment use different systems developed several years ago. For example, the NTSC system in the USA was introduced in 1953 with 525 lines, the SECAM system in France was introduced in the early 1960s with 625 lines, and the PAL system in West Germany and the United Kingdom was also introduced in the early 1960s with 625 lines. Now, over 20 years later, the latest available technology makes it possible for a new system to produce a much-improved picture quality similar to that now provided by the wide-screen cinema three-dimensional colour movies. The problem is to get a system underway that will be acceptable internationally, perhaps with a view to a common DBS system, when in fact development is progressing independently in several countries. A common standard would obviously be a big advantage for future DBS services, and it would seem to be a natural progression to utilize the undoubted merits of digital techniques and the availability of very-large-scale integrated (VLSI) circuit components.

An HDTV analog system has been developed in Japan by the public broadcasting organization NHK and industry which uses 1125 lines at 60 fields per second, with 2:1 interlacing and an aspect ratio of 5:3, compared with the 625 lines and aspect ratio of 4:3 of most existing standards, and a much larger video bandwidth.

One view is that an international standard system should be based on this Japanese one that has already been demonstrated, but another view is that the current system should be enhanced with available technology to produce an acceptable improved performance. Another important consideration is whether a new standard should be compatible with the existing standard, so that on DBS services for example the picture could be received on a conventional small screen or on a high-definition large screen at the choice of the viewer.

This has been the possible future achievement of developments by the IBA in the UK, where a coding system has emerged called C-MAC (Multiplexed Analog Components), with the luminance and colour-difference components being kept separate. The system has been recommended by the EBU as a European standard.

15 Introduction to Optic Fibre Systems

Introduction

A brief introduction was given in Chapter 6 to the use of glass or plastic fibres as an alternative medium to metallic conductors for the transmission of telecommunication systems.

The fibre line acts as a dielectric waveguide to carry light energy which is modulated by an electrical information signal that can be analog or digital in nature. Optic fibre lines give a number of advantages over the use of conventional conducting lines carrying electromagnetic waves having much lower frequencies than light waves. The advantages can be summarized as follows:

a   Smaller lightweight cables, with a smaller bending radius, occupy less space in ducts; ideal for ships and planes where space is at a premium, with fewer men needed for handling.

b   Very large bandwidth available; speech, data and video signals can be transmitted simultaneously by high-capacity systems.

c   Ability to handle natural growth in capacity demand.

d   Low-loss longer sections between repeaters and regenerators.

e   High reliability and long life.

f   Freedom from electromagnetic interference; can be used in noisy electrical situations.

g   Non-inductive and non-conductive; no radiation and interference on other circuits and systems.

h   Continually improving technology is constantly reducing cost and producing more efficient devices and systems.

i   Greater security, since it is very difficult to tap into a fibre cable.

Refraction of a Light Wave

When a light wave passes from one material to another, the direction of travel will be changed. This is called **refraction** and is simply demonstrated by observing that a straight stick appears to bend when it is partly immersed in water. When a light wave meets the junction of air and another material, such as water or glass, at an angle of incidence ϕ_i, then the angle of refraction ϕ_r at which it leaves the junction is different from

Fig. 15.1 Refraction of light by two different materials

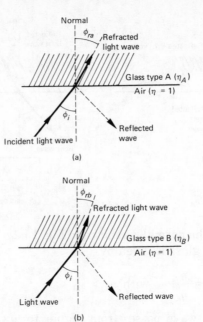

(a)

(b)

the angle of incidence. The principle of refraction of light in this way is illustrated in Fig. 15.1.

In Fig. 15.1, using air and either of two different types of glass for example, the line drawn perpendicular to the point at which the light wave meets the junction of the materials is called the *normal* line. The angle of incidence ϕ_i is the same in Fig. 15.1a and b, but the angles of refraction are different, being ϕ_{ra} and ϕ_{rb} for glass types A and B respectively.

The ratio $\sin \phi_i / \sin \phi_r$ is a *constant* for any two materials and is called the **refractive index** η for the two materials in contact with each other. Since air, with unity refractive index, is used as one of the materials in Fig. 15.1a and b, the ratios are called the *absolute refractive index* of the glass types A and B. So from Fig. 15.1a, the absolute refractive index of glass A is given by

$$\eta_A = \frac{\sin \phi_i}{\sin \phi_{ra}}$$

and is clearly a smaller value than that for glass B which, from Fig. 15.1b, is given by

$$\eta_B = \frac{\sin \phi_i}{\sin \phi_{rb}}$$

since $\phi_{ra} > \phi_{rb}$.

If the two types of glass are now placed in direct contact with each other, the refraction of a light wave will depend on which way it is travelling, as illustrated in Fig. 15.2. A light wave passing from a material of *lower* absolute refractive index to a material of *higher* absolute refractive index

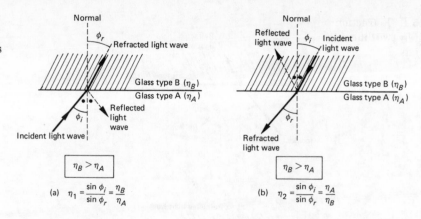

Fig. 15.2 Reciprocal refraction of light by two different materials

(a) $\eta_1 = \dfrac{\sin \phi_i}{\sin \phi_r} = \dfrac{\eta_B}{\eta_A}$

(b) $\eta_2 = \dfrac{\sin \phi_i}{\sin \phi_r} = \dfrac{\eta_A}{\eta_B}$

is bent *towards* the normal, as shown in Fig. 15.2a. The refractive index of the two types of glass is given by

$$\eta_1 = \frac{\sin \phi_i}{\sin \phi_r} = \frac{\eta_B}{\eta_A} \quad \text{with } \eta_B > \eta_A$$

But if the light wave passes from higher to lower refractive index as shown in Fig. 15.2b, it is bent *away* from the normal. The refractive index of the two types of glass is now given by

$$\eta_2 = \frac{\sin \phi_i}{\sin \phi_r} = \frac{\eta_A}{\eta_B} \quad \text{with } \eta_B > \eta_A$$

Clearly, η_2 is greater than η_1. So, there are two possible values of refractive index for these two materials, depending on the direction of travel of the light wave.

$$\text{Since} \quad \eta_1 = \frac{\eta_B}{\eta_A} \quad \text{and} \quad \eta_2 = \frac{\eta_A}{\eta_B} \quad \text{then} \quad \eta_1 = \frac{1}{\eta_2}$$

This reciprocal relationship occurs for any two different materials.

In Fig. 15.2a and b, most of the light energy is refracted, but some is *reflected* from the junction at an angle equal to the angle of incidence ϕ_i, as indicated.

In Fig. 15.2b, with light passing into a material of *lower* absolute refractive index, if the angle of incidence ϕ_i is increased, the angle of refraction ϕ_r is also increased. When the angle of refraction approaches 90°, the light wave does not pass into glass A but will be *totally* reflected, as illustrated in Fig. 15.3.

The angle of incidence at which **total reflection** first occurs is called the *critical angle* ϕ_c for the two types of glass. Light waves incident at angles *greater* than ϕ_c will also be totally reflected.

Now, since in Fig. 15.2b,

$$\frac{\sin \phi_i}{\sin \phi_r} = \frac{\eta_A}{\eta_B}, \quad \text{then}$$

$$\sin \phi_i = \frac{\eta_A}{\eta_B} \sin \phi_r$$

Fig. 15.3 Total reflection of light by two different materials

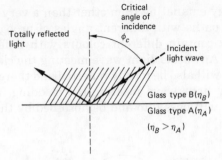

So, in Fig. 15.3, for the critical angle of incidence ϕ_c,

$$\phi_r = 90° \quad \text{and} \quad \sin 90° = 1$$

$$\therefore \quad \sin \phi_c = \frac{\eta_A}{\eta_B} \quad \text{and} \quad \phi_c = \sin^{-1}\frac{\eta_A}{\eta_B}$$

Now, η_B is always slightly greater than η_A, so

$$\sin \phi_c \text{ will be less than 1}$$
and $\quad \phi_c$ will be less than 90°

For example, in a typical fibre cable, $\eta_A = 1.45$ and $\eta_B = 1.47$.

$$\therefore \quad \sin \phi_c = \frac{1.45}{1.47} = 0.9864 \quad \therefore \quad \phi_c = 80°32'$$

Propagation in an Optic Fibre

In Chapter 6, Fig. 6.6 illustrated the basic construction of an optic fibre line. It consists of a glass core, with a certain absolute refractive index η_B, totally enclosed by a glass cladding having an absolute refractive index η_A lower than that of the core.

A longitudinal cross-section of a fibre is shown in Fig. 15.4. A light wave travelling along the core that meets the cladding at the critical angle of

Fig. 15.4 Propagation of light along a glass-fibre core

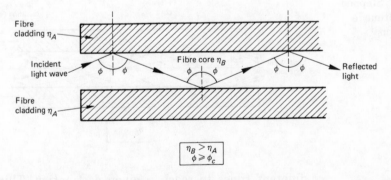

incidence ϕ_c will be totally reflected. It then meets the opposite surface of the cladding again at the critical angle ϕ_c and so is again totally reflected. The light wave is therefore propagated along the core by a series of total reflections from the cladding.

Light energy emanating from other than a very tiny point source will have several paths with different angles of propagation, and moreover will probably contain different colours with different frequencies and wavelengths. Any other light wave meeting the cladding at or above the critical value will also be totally reflected and therefore propagated along the core. Any light wave meeting the cladding at an angle *below* the critical value will pass into and be absorbed by the cladding.

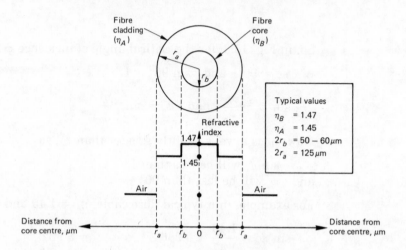

Fig. 15.5 Refractive index profile for stepped-index multimode fibre

There is clearly a sudden change of absolute refractive index at the junction of the core and the cladding, which is illustrated by the refractive index profile diagram of Fig. 15.5. The fibre is therefore called a *stepped-index fibre*, and since we have seen that it can carry many different light waves, then **stepped-index multimode** propagation occurs as illustrated in Fig. 15.6. The various light waves travelling along the core clearly have propagation paths of different lengths, and so will take

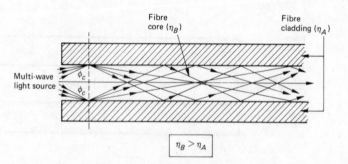

Fig. 15.6 Stepped-index multimode propagation

different times to reach a given destination. This produces distortion known as *transit-time dispersion*, which introduces an upper limit on the *rate* at which the light can be modulated by an analog or digital electrical signal. Otherwise the variations or successive pulses of light will merge into each other, and cause distortion of the information being carried.

Fig. 15.7 Refractive index profile for stepped-index monomode fibre

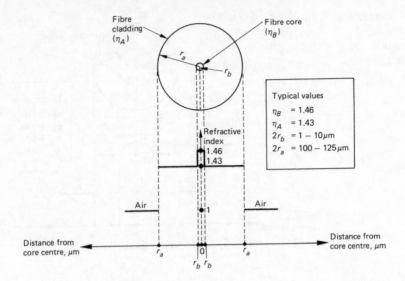

Fibre cladding (η_A)

Fibre core (η_B)

r_a

r_b

Typical values
η_B = 1.46
η_A = 1.43
$2r_b = 1 - 10\mu m$
$2r_a = 100 - 125\mu m$

Refractive index
1.46
1.43

Air

Air

1

Distance from core centre, μm

Distance from core centre, μm

r_a

r_b r_b

0

r_a

Fig. 15.8 Stepped-index monomode propagation

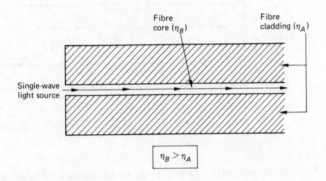

Fibre core (η_B)

Fibre cladding (η_A)

Single-wave light source

$\eta_B > \eta_A$

Stepped-index multimode fibres are therefore used generally for data links requiring relatively low bandwidth and capacity.

The problems of transit-time dispersion can be overcome by making the core very thin, so that its diameter is of the same order as the wavelength of the light wave to be propagated, as illustrated in Fig. 15.7. The resultant propagation of a single light wave is shown in Fig. 15.8. This type of fibre is called a **stepped-index monomode** fibre. However, the use of very thin cores creates mechanical difficulties in the manufacture, handling and jointing of the fibres, and is therefore very expensive. The use of stepped-index monomode fibres is generally limited to very-high-capacity large-bandwidth systems, such as undersea cables where the expense is justified by the high return of earned income.

For most intermediate bandwidth and capacity systems at present in operation, a less-expensive method of overcoming transit-time dispersion is used. The core is made of a material which has a *graded* absolute refractive instead of a fixed value, being highest at the core centre and gradually decreasing towards its outer edge where it meets the cladding. This is called a **graded-index** fibre, with a profile as illustrated in Fig. 15.9.

Fig. 15.9 Refractive index profile for graded-index multimode fibre

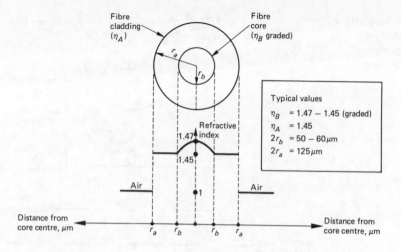

Fig. 15.10 Graded-index multimode propagation

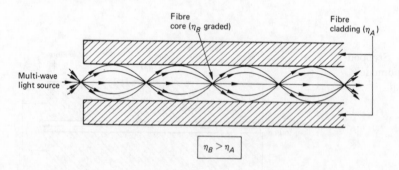

The propagation of multi-wave light energy is shown in Fig. 15.10, with the individual waves being gradually refracted in the graded-index core instead of being reflected by the cladding. Waves travelling at different incident angles will travel different distances from the horizontal central axis before being refracted back to re-cross the central axis.

Although waves with a large angle of incidence travel further than those with smaller angles, the decrease of refractive index also allows a higher velocity of propagation of energy. As a result, all waves will reach a given point along the fibre at virtually the same time, thus greatly reducing transit-time dispersion. This is called **graded-index multimode propagation**.

In all three types of fibre described above, the thickness of the cladding must be several wavelengths to prevent light energy losses due to absorption and scattering.

Optic Fibre Telecommunication Systems

Information in electrical form can be speech or video signals or coded digital signal pulses as with pcm systems. Either can be used to modulate light energy produced from a source such as a laser diode or light-emitting diode.

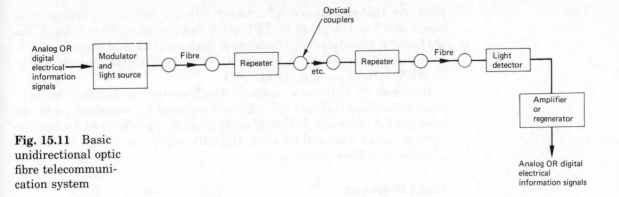

Fig. 15.11 Basic unidirectional optic fibre telecommunication system

The modulated light is then transmitted along an optic fibre line until regeneration is required by a repeater. At the distant end, a detector such as a semiconductor photodiode converts the modulated light energy back into the original analog or digital electrical signals (see Fig. 15.11).

Light Sources

The light source shown in Fig. 15.11 must deliver light energy at the required wavelength or wavelengths, must respond to modulation by relatively low-power digital electrical information signals, and must deliver adequate light power to be launched along the fibre. Semiconductor diodes constructed from particular materials will emit light energy when conducting under the control of forward biasing potential, and the output light energy can be varied or switched on and off very rapidly.

Typical materials are gallium arsenide (GaAs), gallium aluminium arsenide (GaAlAs), and indium gallium arsenide phosphide (InGaAsP). Some of these materials, of course, were used as impurity or doping agents in early silicon and germanium semiconductors. **Light-emitting diodes** (LEDs) and **injection-laser diodes** (ILDs) are the two main sources at present in use. Their physical dimensions are compatible with those of optic fibres, they can emit visible and infra-red light at wavelengths that correspond with the low-loss characteristics of optic fibres, and their light energy outputs can be rapidly controlled, so they are easy to modulate. Furthermore they have the reliability and long life that is typical of solid-state components.

LEDs and ILDs have certain similarities, but they also have important differences that influence the choice for particular applications. LED sources emit a relatively wide beam of light with a fairly large spectral bandwidth, so they are suitable for stepped-index multimode propagation. ILD sources radiate a much narrower beam of light with a much narrower spectral bandwidth. The presence of a resonant cavity produces light output at a single wavelength, so ILD sources are compatible with stepped-index monomode fibres.

The width of the light beam and the spectral bandwidth of a source determine the amount of light power that can be launched into a fibre, and the amount of dispersion that will occur in the working bandwidth of the fibre. An ILD can launch light power into the fibre in an approximate range of 0.5 to 5 mW, but an LED is limited to about 0.05 to 0.5 mW. An ILD can be modulated at a faster rate than an LED, and it has a significantly greater modulated bandwidth of over 500 MHz compared with about 50 MHz maximum for an LED.

However, an ILD needs some feedback control to stabilize the light output. GaAlAs ILD and LED devices are used for wavelengths in the range of 0.8 to 0.9 μm. InGaAsP devices are being developed for applications around 1.3 μm and 1.6 μm in the infra-red region where the loss and dispersion in fibres are very low.

Light Detectors

The function of the light detector shown in Fig. 15.11 is to detect the modulated light signal travelling along the fibre and to convert the variations or pulses back into the original analog or digital information signals that modulated the light source at the sending end. A reverse-biased semiconductor **photodiode** will conduct when subjected to light energy, with an output current that is virtually proportional to the amount of light energy impinging on it. If the output current is passed through a load resistor, an output voltage is obtained.

There are two main types of semiconductor photodiode, the **PIN** diode and the **avalanche** photodiode (APD). The PIN diode has an intrinsic semiconductor material placed between the p-type and n-type regions, and has a conversion efficiency from input light photons to output electrons of less than unity. The APD however provides an effective conversion gain by charge carrier multiplication resulting in the production of multiple electrons per photon. So, the choice between PIN diode and APD depends largely on the required receiver sensitivity and acceptable cost.

Silicon PIN diodes and APDs are used for wavelengths of up to 1.1 μm, but germanium types can be used up to about 1.8 μm. More recently, InP, GaInAs and GaInAsP devices have been introduced for use at the longer wavelengths where silicon types are unsuitable.

Repeaters

The repeaters shown in Fig. 15.11 are needed at regular intervals along the fibre line so that the modulated light energy does not fall below a detectable level due to losses. The repeater consists basically of a PIN diode or APD detector, an electronic pulse regenerator, and an LED or ILD light source to send the modulated light signals on to the next section of fibre line.

Development of Optic Fibre Systems

In the UK the first optic fibre telecommunication link was installed in 1980, and by 1984 an extensive network had been established. It is estimated that about one half of all long-distance telephone routes will use optic fibre links before the end of the decade. It is unlikely that any new coaxial cable links will be installed after 1985. Installation costs of the optic fibre cables will be low because existing cable ducts and terminal buildings will be available. A fibre cable laid under the English Channel to Belgium will be operational by the mid-1980s capable of carrying up to 1200 speech channels.

In the USA an optic fibre system was installed between Washington and New York in 1983 with a bandwidth capacity of over 1300 speech channels, and many similar systems have followed.

In Japan, a local cable TV system was installed in 1978 using optic fibres in a star distribution network. This system employs analog intensity-modulation techniques as opposed to the digital techniques adopted by most other optic fibre systems introduced elsewhere.

In France, an experimental small optic fibre network has been installed to replace the normal local line distribution cable network serving about 5000 subscribers. It is hoped to demonstrate that, as well as improving the quality of the telephone service, additional services such as tv, vhf radio, data and facsimile can also be provided on a single network.

An interesting development of optical communication is the use of infra-red light direct over short distances without the use of fibres. One example in Switzerland is an inter-station system in a large office controlled by a central light source/detector installed in the ceiling. Also, British Telecom have plans to set up a local telephone distribution network in a small area of London using roof-mounted infra-red devices. So there are many varied and exciting possibilities for the rapid development of optic fibre systems in the near future.

16 Introduction to Mobile Radio Systems

Introduction

The second world war (1939–1945) saw many rapid technological developments for military purposes which also had significant impact in the ensuing peace. One such development was the use of mobile radio services, that is radio communication between central control stations and moving stations such as motor vehicles, tanks, aeroplanes and fighting personnel. Clearly such mobile stations are only capable of carrying aerials of relatively small physical size.

The efficiency of an aerial is directly related to the wavelength of the radio frequency employed for the transmission. It was shown in Chapter 1 (page 8) that wavelength is the reciprocal of frequency, so it follows that efficient aerials of small physical size require high working frequencies. The urgent need for mobile communication in wartime therefore prompted rapid advances in component and circuit techniques for using very high frequencies (vhf) in the broad range 30–300 MHz (relative wavelengths 10–1 metres).

At these frequencies, it is possible to keep mobile aerials to an acceptable physical size, and it is possible, where necessary, to design aerial arrays having significant directivity, thus improving security of services compared with the use of omnidirectional "broadcast" aerials.

When the war ended there was universal realization that the use of mobile radio communication was attractive for many applications, such as for police, ambulance and fire-fighting services, and for doctors, taxis and a host of other private business uses. On a national and international basis therefore, certain radio frequency bands were allocated for mobile radio services, for example in the vhf range around 80 and ,160 MHz initially. The appropriate bands were divided into a number of equispaced channels, and the allocation and licensing of channel frequencies was undertaken by a controlling authority in each country. In the United Kingdom the authority was originally the Post Office, but later to be taken over by the Home Office. Restriction of radiated power to a specified maximum meant that there was a limited area that could be covered from a base station, as illustrated in Fig. 16.1.

Fig. 16.1 Typical mobile radio system service area

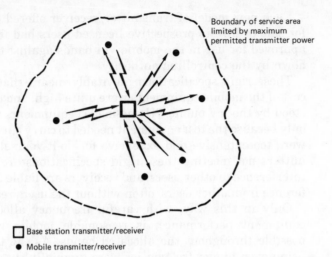

Boundary of service area limited by maximum permitted transmitter power

☐ Base station transmitter/receiver

● Mobile transmitter/receiver

The demand for these private and official mobile radio services rapidly increased, far outnumbering the number of channels available in the allocated bands. The original channel spacing in the UK was 100 kHz, but this was twice reduced, first to 50 kHz and then to 25 kHz to increase the number of channels available. Despite this, the demand for services was such that allocation of exclusive channels to each user was impossible, and channel sharing was necessary.

Furthermore, a number of problems became evident as the user population increased rapidly, and it soon became apparent that, to provide a reasonably satisfactory service to all users, frequency allocation had to be done very carefully in any one area, and that the technical performance of the **base station** and **mobile transmitter/receiver** equipment had to be carefully specified and enforced.

For the transmitter, the specification mainly covered maximum permitted radiated power, accuracy and stability of the generated channel carrier frequency, degree of modulation, and out-of-channel interference radiation (due to harmonics and sub-harmonics of carrier frequency, and to overmodulation). It should be added here that, to generate an accurate and stable carrier frequency at vhf, it was usual to use a low-frequency crystal-controlled oscillator followed by appropriate frequency multipliers to produce the allocated and licensed carrier frequency.

Amplitude, frequency or phase modulation was permitted originally at the choice of the licensed user and equipment manufacturers. For the receiver, the problems were even more difficult, surprising though this may seem, and rigid specifications were needed for sensitivity, desensitization (blocking), adjacent channel selectivity, intermodulation and cross-modulation interference, spurious responses, and accuracy and stability of the crystal-controlled local oscillator.

For transmitter and receiver, the specifications required tests to be made within a given range of power supply voltage variations (typically ±10%) between a given range of maximum and minimum working temperatures (typically −10 to +40°C).

Each new model of transmitter/receiver offered by any manufacturer for sale or hire to prospective licensed users had to be certified as type-approved for use in the mobile vhf bands against the specifications laid down by the controlling authority.

These rigid specifications inevitably meant that the complexity and cost of the mobile equipments were quite high, and often not at all understood by the lay public. "Home-made" equipments were unacceptable, if only because the test equipment needed to carry out the specification tests were too extensive and expensive for do-it-yourself enthusiasts. Transmitters not meeting these rigid specifications are very liable to cause interference to other users, and receivers are liable to unacceptable interference from other users, often without the users realizing the problems.

Only in this way, with careful frequency allocation and control of equipment performance, is reasonable mobile radio communication possible throughout the allocated bands. As the technical problems of using even higher frequencies were gradually overcome, a further band was allocated to mobile radio services in the uhf range around 450 MHz.

Two-frequency Operation

It was mentioned in Chapter 3 (page 22) that mobile radio systems and other applications of two-way radio communication systems use different frequencies for the two directions of communication. The reason for this can be explained simply as follows. If a mobile station is transmitting an allocated carrier frequency on channel 1, then for a mobile receiver of any other service in the same area, many consecutive channels are "sterilized" or unusable due to the characteristic called **desensitization** (or *blocking*) caused by the high radiated power on channel 1 overriding the selectivity of a receiver. So, in any one area, only a small number of the total available channels can be used. This is illustrated in Fig. 16.2.

In Fig. 16.2, with single-frequency working (base and mobile using the same transmitter frequency), f_2 from mobile 2 to base 2 must be separated from f_1 by many channels in order for mobile 1 receiver to operate satisfactorily with signals from base 1 without desensitization due to mobile 2. So, single-frequency working results in inefficient use of the available frequency band in any one area.

Fig. 16.2 The problem of overlapping service areas

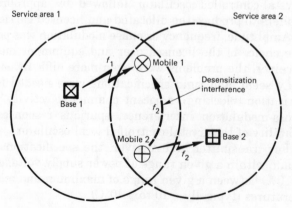

Service area 1

Service area 2

f_1

Mobile 1

Desensitization interference

Base 1

f_2

Mobile 2

f_2

Base 2

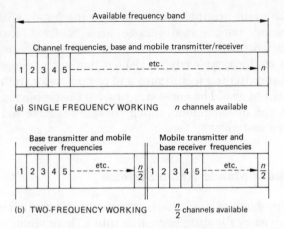

Fig. 16.3 Comparison of single-frequency and two-frequency working

Available frequency band

Channel frequencies, base and mobile transmitter/receiver

| 1 | 2 | 3 | 4 | 5 | etc. | n |

(a) SINGLE FREQUENCY WORKING n channels available

Base transmitter and mobile receiver frequencies

Mobile transmitter and base receiver frequencies

| 1 | 2 | 3 | 4 | 5 | etc. | $\frac{n}{2}$ | 1 | 2 | 3 | 4 | 5 | etc. | $\frac{n}{2}$ |

(b) TWO-FREQUENCY WORKING $\frac{n}{2}$ channels available

With two-frequency working, the allocated band is divided into two halves: one half is used for base station transmitting channel frequencies, the other half is used for the corresponding mobile transmitter frequencies. This is illustrated in Fig. 16.3, and f_2 would be separated from f_1 by an adequate amount even for adjacent channels.

This two-frequency per channel arrangement may seem at first glance to be wasteful of the available band, since the number of channels is only half that with single-frequency working. But in fact more channels can be used in any one area because the problem of receiver desensitization is less severe.

Links to the Public Switched Telephone Network

As the use of private and official vhf/uhf mobile radio services rapidly expanded it became apparent that many wide-ranging benefits would result if the normal public telephone network could be extended to mobile users, both in vehicles and for hand-held personal radios. The first attempt in the UK was made in the early 1960s by the Post Office (later to become British Telecom) with the introduction of a pilot scheme in South Lancashire. A base radio station was installed near Horwich, with the system controlled by telephone operators in a nearby exchange. As with any fixed base station there was a limited service area, within which properly licensed and equipped vehicles had access to a calling channel to the controlling exchange. The telephone operator set up the required call to the pstn and connected it to the vehicle concerned on one of a small number of hand-switched working channels. As a result of this pilot scheme a similar one was later introduced in London with, eventually, five base stations and more working channels than in the South Lancashire scheme.

However, there was a limited service area and insufficient channels to meet the growing demand by prospective subscribers, and it was inevitable that thoughts were directed to the need for a more flexible system with much greater capacity. This need of course was arising everywhere, and indeed a survey in Europe indicated that approximately 30% of the working population are mobile for a part or whole of each working day.

The main original problems were the hand switching of channel frequencies, the limited service area of fixed base stations, and the inefficient use of channels because of interference from the high-power base stations. As early as 1947 it had been suggested in the USA that digital switching techniques and computer-controlled automatic channel switching would be necessary to solve the problems. This proved to be the case and developments along these lines have led to present plans for future systems.

Cellular Mobile Radio Systems

The problem of single high-power base stations in any one particular area is solved by dividing each area into a "honeycomb" of small sub-areas or **cells**. Each cell has one low-power omni-directional base transmitting aerial (or, in some cases, up to 2 or 3 directional aerials at appropriate locations in the cell).

These base stations are connected by land-lines to mobile telephone exchanges (MTX), strategically placed as appropriate, with access to the pstn. Because of the low power employed, channel frequencies can be used again for non-adjacent cells in any area, and in other areas, giving optimum use of all possible available channels. This is illustrated in Fig. 16.4.

As a mobile moves about from one cell to another within an area, and from one area to another, the controlling mobile station monitors all available channel signals, detects the strongest signal in any position, and arranges for the appropriate channel to be automatically selected by computer control. Change of channel can be made in approximately 300 msec and is said to be undetectable in conversation.

Although the service areas of adjacent low-power transmitters obviously overlap as shown in Fig. 16.4, the fact that the computer-controlled equipment can select the strongest available signal has led to the representation of a cellular system by a honeycomb effect as illustrated in Fig. 16.5.

The main advantages of cellular radio systems are economy, efficiency, flexibility, and great potential of facilities. For example, the size of cells can be changed as growth occurs. Small cells may have an area as small as $1 \, \text{km}^2$, whereas large cells may have $50 \, \text{km}^2$. Each cell may require anything from 2 to 50 channels, and it is estimated that each channel can cater for approximately 30 subscribers. The mobile telephone exchanges can also be increased in capacity as demand increases.

Typical systems In Europe, frequencies have been allocated in the uhf band in the range 890–989 MHz, with two-frequency working and 25 kHz channel spacing.

In the USA, the allocated frequency bands are approximately 70 MHz lower than in Europe, with 30 kHz channel spacing.

Four Scandinavian countries have collaborated to set up the Nordic Mobile Telephone (NMT) network, which has at present about 100 000 subscribers, with 400 000 forecast by 1991.

Fig. 16.4 Simple principle of cellular radio system

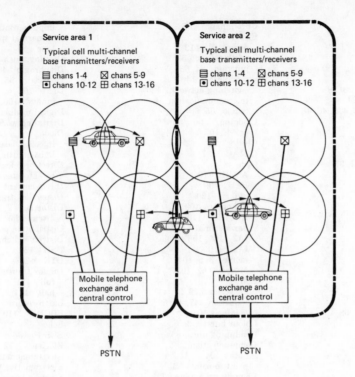

Fig. 16.5 "Honeycomb" cellular system representation

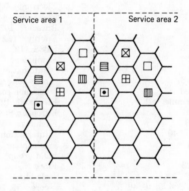

In the UK, two systems have been developed, which commenced operation around March 1985. One system is called "Cellnet", developed by British Telecom and Securicor. The second is the Racal "Vodafone" system. Both these systems conform to standards entitled TACS (Total Access Communication System) which is based on the AMPS (Advanced Mobile Phone System) developed in the USA.

In Canada, a system is already working in the 400 MHz band, with a further system scheduled soon around 800 MHz.

Systems have also been set up in Spain, Netherlands, Luxembourg, Saudi-Arabia, Oman, Malaysia, Tunisia, Austria, Bahrain and Korea, with plans progressing in Australia and Hong Kong.